The
Language of
Science

The Language of Science

MH Orr
Department of English, UNISA

CJH Schutte
Dean of the Faculty of Science, UNISA

Butterworths
Durban

© 1992
Butterworths/MH van Zyl and CJH Schutte

Butterworth Publishers (Pty) Ltd
Reg No 91/05175/07

ISBN 0 409 11165 1

Reprinted 1994

Durban
8 Walter Place, Waterval Park
Mayville 4091

Johannesburg
108 Elizabeth Avenue
Parkmore 2196

Pretoria
Suite 301, 270 Main Street
Waterkloof 0181

Cape Town
3 Gardens Business Village, Hope Street
Cape Town 8001

Front cover: wood-block engraving from the 1736 Luther Bible
showing God and the animals (natural as well as mythological),
to whom Adam gave names.

Typeset by Digma Publications
Printed and bound by Interpak Natal Pietermaritzburg

Preface

Readers in an academic environment often feel overwhelmed and intimidated by the complexity of the texts with which they have to deal in science courses. English classes at school seldom equip readers with the unique reading skills that the sciences demand, and few are fluent in the unique 'languages' of the sciences (or indeed, even aware that such special 'languages' exist). These difficulties are exacerbated for those for whom English is not a first language. *The Language of Science* arose out of a desire to meet the needs of students, scientists and lecturers (in English or in Science) facing this challenge.

The book is an interdisciplinary and co-operative effort between language teaching and science, and the two apparent polarities are reflected in the style and the content. Two obvious sections emerge: Chapters 1 to 8, and Chapters 9 to 13.

The first section is a language skills package for students and practitioners of any science. Readers will find that these language skills differ from the content of conventional language classes, in that they focus specifically on the English used in science. Initial investigation by the authors indicated a number of characteristics unique to science texts, and hence a range of skills adapted to these. It is these particular language skills that this text addresses.

Chapters 1 to 8 provide the reader with a basic 'tool kit' for tackling the reading of science textbooks. Chapter 1 starts on the micro level of words and their functions, and successive chapters guide the reader through the assembling and constructing of ideas and information in sentences and paragraphs. We discuss the basic types of paragraph common to science texts and provide reading and study strategies for dealing with these. Chapter 8 provides some sophisticated apparatus for the 'toolkit' in the form of thinking skills – strategies for managing and manipulating information.

The second part of the book, comprising Chapters 9 to 13, deals with some elementary aspects of the language encountered in biology, mathematics, chemistry and physics texts. Matters such as the following are considered:

☐ etymologies of words
☐ the construction of compound nouns, verbs and adjectives
☐ names and terms constructed according to some logical nomenclature schemes
☐ symbols, conventions, abbreviations, formulae, equations, units and directions
☐ definitions and their interrelationships as permitted and required by any system of definitions
☐ visual information, such as graphs, tables, colour codes, abstract figures representing reality, graphical definitions, unspoken visual conventions, etc.

The treatment is intended to be of direct practical use to the reader, to enable him or her to understand the more elementary type of textual material found in the sciences. The more fundamental, philosophical and linguistic aspects of the subject are thus not addressed. We do not try to teach science through the examples and problems given in the text, but focus on the language, conventions and underlying structure.

There are various 'reading paths' through the book. The reader should systematically work through Chapters 1 to 8 and Chapter 13 (visual information). Chapters 9 to 12 may then be studied in any sequence or combination as dictated by the interests of the reader, for instance:

☐ only Chapter 9 (biology), or
☐ only Chapter 10 (mathematics), or
☐ Chapter 11 (chemistry) and Chapter 12 (physics and physical chemistry), or
☐ Chapter 10 (mathematics) and Chapter 12 (physics and physical chemistry).

We wish to express our thanks to colleagues who generously gave their time and expertise to read various drafts of chapters: Prof SJR Vorster and Prof WA Labuschagne (Mathematics), Prof G McGillivray and Prof SO Paul (Chemistry), Prof EC Reynhardt (Physics and Visual Information), Prof OG Bosch and Prof AJ Reyneke (Biology). Of course, any infelicities or inaccuracies remaining are our responsibility.

We trust that the reader will learn as much about language in science from studying this book as we have learnt in writing it.

MH Orr and CJH Schutte

Contents

Introduction: Reading Skills

If you were not able to read, you would not have this book in your hands right now. So what is the point of a section on reading skills? The point is, partly, to put a drawing pin on your seat – to galvanise you out of a complacent, passive attitude towards reading. You have probably become so used to reading by now, that you may see it as merely a dutiful running of the eyes over x pages of text, at the end of which you emerge with y understanding. Or, perhaps, reading (particularly in English) has always been problematic for you, and you tend to plod grimly through pages of text, battling with unfamiliar vocabulary and complicated sentences, feeling like you're wading through porridge. Either way, your attitudes towards reading may have become fossilised and unproductive – mindsets that the authors of this book hope to transform.

One of the aims of this book is to make you a more efficient reader of scientific prose, and to that end you will find – in the chapters that follow – a toolkit of strategies; hints on how to summarise texts for study purposes; explanations of the patterns, structures, and content common to scientific texts; and general guidelines on processing complex information. First, however, it is useful to offer you some insight into what constitutes effective and efficient reading.

(1) *Reading is dynamic*

Reading is not a passive skill – the text will not automatically print straight from your eyes into your knowledge structures. You have to wrest meaning from the text by being a dynamic reader, creating and integrating meaning as you read.

(2) *Reading is goal-directed*

Always set goals before you start to read. Think about what you are reading the text for, what your desired outcome is. *Why* you are reading the text will determine *how* you read it.

(3) *Reading differs according to the reader's goals*

You should not read all texts in the same way. This should be obvious – one does not read the telephone directory in the same way that one reads a letter from one's lover. However, many students feel that all academic texts have to be read identically – with dogged determination, and strict attention to every single word. In fact, a range of academic reading skills exists, and flexible readers know when to use each skill, thus making their reading faster and more efficient.

☐ Reading Skill 1: Scanning

Definition

> Scanning is a rapid type of reading done to obtain a specific fact, or specific piece of information.

An everyday example of scanning is the type of reading one does to look up a telephone number. You run your eyes rapidly down the column of names until they are 'hooked' by the right name, whereupon you stop and read carefully. Scanning does *not* involve you in reading every word.

As a student, you will need to use scanning for the following tasks, amongst others:

☐ using a dictionary or glossary
☐ using an index
☐ looking up important dates in the university calendar
☐ using tables
☐ finding books in the library
☐ looking for articles in a journal

How to scan

Successful scanning depends on two things: speed and visualisation of the piece of information you are looking for.

☐ Use alphabetical order.
☐ Use numerical order.
☐ Think about what form the information will take.

    – a word or a number?
    – is the word capitalised?
    – what letter does the word begin with?
    – is it a key phrase? how many words?

☐ Picture the item in your mind before starting to scan.

☐ Reading Skill 2: Skimming

Definition

> Skimming is a rapid type of reading done to obtain the general idea of a paragraph or piece of text.

People often skim an article in a magazine while waiting in the queue at a supermarket till. You skim by reading the title, looking at the photographs or illustrations, reading the captions and bold headings, and letting your eyes swoop over the article, coming to rest on key words and phrases here and there ... all in all getting a fairly general idea of what the article is about. (And, also, perhaps, helping you to decide if you want to buy the magazine to read more closely at home!)

One usually skims with a general question in mind such as: 'What is this all about? What is the author talking about?' In an academic context, one generally skims when there is no time to read carefully (for example, skimming a prescribed chapter just before a lecture on it) or when one is trying to decide whether an article or passage needs more careful reading. In the latter case, one might be reading with a slightly more specific question in mind, such as – 'Does this article offer me any new information about Heisenberg's Uncertainty Principle, or does it cover stuff I already know?'

How to skim

☐ Read rapidly (almost as fast as scanning).
☐ Do not read every word.
☐ Let your eyes 'hook' on key phrases (nouns, verbs, and proper nouns are important).
☐ Formulate questions before you begin, and read with these in mind.

☐ Reading Skill 3: Comprehensive Reading

Definition

> Comprehensive reading is a systematic, fairly slow type of reading done to arrive at a thorough, detailed understanding of a piece of text.

Instructions for connecting a signal decoder to your video recorder and television set are the type of everyday text on which one practises comprehensive reading – otherwise chaos and frustration result. In the academic world, you will find that a great deal of your material demands comprehensive reading – textbooks (including this one), instructions on how to get a computer program up and running, articles in academic journals, lecture notes.

How to read comprehensively

☐ Read slowly. (You will often have already skimmed the text before you start comprehensive reading.)
☐ Read carefully in order to get the total message (main idea, supporting details, examples and illustrations).
☐ Backtrack and re-read if you lose the thread of the argument.

One stumbling block to successful comprehensive reading may be difficult **vocabulary.** If you encounter a word you do not understand, try one of the following strategies:

☐ Use the context as an aid to making an intelligent guess about the meaning.

☐ Look it up in a dictionary or the glossary of the textbook you are reading.

☐ Ignore it unless the word is crucial to your understanding. You can often understand the gist of a text without necessarily knowing the meaning of every single word in it.

Chapters 1 to 7 of this textbook will offer you additional guidelines and strategies for this type of reading – use them as a 'Comprehensive Reading Toolkit'.

☐ Reading Skill 4: Thoughtful Reading

Definition

> Thoughtful reading is a slow, intensive interaction with text, in which the reader is required to manage and manipulate information in order to arrive at complex insights or solutions.

This is the most difficult reading skill to master. It is the type of reading required by mathematical proofs, for example. Ultimately, your success as a student in the sciences depends on your ability to organise and process complex information by reading thoughtfully and thinking logically.

How to read thought-fully

☐ Read slowly and carefully.

☐ Make sure you have a grip on the facts.
(Can you list them?)

☐ Think about the *implications* of the facts.
On what evidence or assumptions are they based?
What conclusions or deductions do they lead to?

☐ Ask yourself questions as you read.
(Who ... ? Where ... ? When ... ? Why ... ? How ... ? What ... ?)

☐ Try to understand abstract ideas by forming pictures in your head as you read, or by thinking of practical, real-life illustrations of the concepts.

☐ Do not panic if you get lost or confused.
Think systematically and positively.

Chapter 8 offers you practical tactics and strategies to enhance your logical thinking skills – study it carefully.

1

Words

The English language has some 500 000 words and 300 000 technical terms – an overwhelming number to learn and master! However, no one knows all these words, and most of us get by with a relatively small vocabulary. More important, perhaps, than learning lists of words, is learning the principles governing their formation and use, and a few basic strategies for dealing with unfamiliar words. In this chapter, we will be looking at the basic characteristics of the major word classes – nouns, verbs, adjectives, and adverbs.

☐ Nouns

Definition

> A noun is a word naming a person, place, thing, phenomenon, idea, quality, or condition.

Examples

The italicised words in the following paragraphs are all nouns:

> When a heavy *body* such as a *stone* or a *piece* of *iron* is suspended on a *string* and totally immersed in *water*, there are three *forces* acting on it: the downward *force* due to the *weight* of the *object*, the *upthrust* due to the displaced *water*, and the *tension* of the supporting *string*.
>
> Allen, J B & H G Widdowson
> *English in Physical Science* Oxford University Press 1979 37

> *Mesohippus* evolved about 35 millions of *years* ago. It had only three *digits* on its *forelimb*, although the fifth was represented by a splint *bone*. The *climate* and the *vegetation* were still similar to those of the *Eocene*, and *Mesohippus* also fed on broad-leaved *plants*. However, by the *middle* of the *Miocene* the *climate* was drier and soft *grasses* replaced the broad-leaved *plants*.
>
> Pearson, Ian
> *English in Biological Science* Oxford University Press 1978 92

Features of nouns

Nouns, as a class of words, have the following features which will help you to identify them:

(1) Nouns provide the answer to the question 'who?' or 'what?' in a sentence.

5

(2) Nouns can be replaced by pronouns (words such as *I, me, you, he, him, she, her, it, we, us, they, them, this, these*), or the words 'thing', 'something' or 'someone'.

☐ *Animals* must obtain *food* – *they* must obtain *it*
☐ When a heavy *body* is suspended on a *string* and totally immersed in *water*... – when a heavy *thing* is suspended on *something* and totally immersed in *something* ...

(3) Nouns can be made into plurals, usually by adding -s or -es.

☐ body – bodies   ☐ stone – stones   ☐ piece – pieces

(4) Nouns can be made into possessives, usually by adding an apostrophe and an -s.

☐ the object's weight   ☐ the string's tension

(5) Nouns can be preceded by words such as *a, an, the, my, each, some, many, this, that,* or another possessive noun or pronoun. Words such as these signal that a noun is coming.

☐ the reason   ☐ this tension
☐ its forelimb   ☐ a string

(6) Nouns can be formed by adding certain suffixes (word-endings) to words. Typical noun suffixes include:

☐ -hood (child + hood = childhood)
☐ -age (mile + age = mileage)
☐ -al (refuse + al = refusal)
☐ -ness (happy + ness = happiness)

Practice session 1

Underline all the nouns in the following paragraphs. (Do not concern yourself so much with the meanings of the words – depending on your area of specialisation, many of the words may be unfamiliar to you. Just try to detect the nouns on the basis of their characteristic features and their function in the sentences.)

(1) A hydraulic press is a device for securing mechanical advantage from the pressures of liquid columns. In its simplest form it consists of a wide cylinder and a narrow cylinder connected by a tube. Both cylinders are filled with water and fitted with a piston.

Allen, J B & H G Widdowson
English in Physical Science Oxford University Press 1979 52

(2) Most species of Euglena possess plastids with chlorophyll inside (ie chloroplasts) and therefore they can photosynthesize. However, all the green species are unable to synthesize at least one organic substance that they need, and they must obtain

continued

continued

these substances osmotrophically. They are therefore partly autotrophic and partly heterotrophic. The colourless species must obviously be fully heterotrophic.

Pearson, Ian
English in Biological Science Oxford University Press 1978 13

(3) Let us consider an example of a process which illustrates the relationship between acids, bases, and salts. If a solution of sodium hydroxide is poured into a beaker containing hydrochloric acid and red litmus, the litmus will turn blue. This shows that the acid has been neutralized and that sodium hydroxide is a base. As a result of the neutralization of the acid by the base, sodium chloride is formed and may be obtained as crystals by evaporation of the solution. Thus the result of adding the base sodium hydroxide to hydrochloric acid is to form a salt, sodium chloride. This substance forms as a result of the metal sodium replacing the hydrogen in the acid.

Allen, J B & H G Widdowson
English in Physical Science Oxford University Press 1979 23

(4) Animals ... must obtain complex organic substances by eating plants or other animals. The reason for this is that they lack chlorophyll. Among these 'other-feeders' or heterotrophs, we distinguish between 'solid-feeders' or phagotrophs, and 'liquid-feeders', or osmotrophs. Whereas phagotrophic organisms take in solid and often living food, osmotrophic ones absorb or suck up liquid food. This is usually from dead and rotting organisms.

Pearson, Ian
English in Biological Science Oxford University Press 1978 13

Check your answers against those at the end of this chapter before continuing. If you made an error, try to determine what went wrong.

Types of nouns

There are four main types of nouns:

(1) Proper nouns are the names of specific people, places, and things. Proper nouns are always written with an initial capital letter.

☐ Mesohippus   ☐ Miocene   ☐ English
☐ Jupiter   ☐ Archimedes   ☐ Heisenberg

(2) Common nouns are the names of ordinary, everyday objects, people, places, or things perceivable by the senses.

☐ stone   ☐ plants   ☐ chlorophyll
☐ cylinder   ☐ sodium   ☐ osmotrophs

(3) Abstract nouns are labels for ideas, qualities, conditions, or phenomena not obviously perceivable to the five concrete senses.

☐ advantage   ☐ theory   ☐ deduction
☐ technology   ☐ hypothesis   ☐ intellect

(4) Collective nouns are the labels for groups of things considered as a unit.

☐ flock   ☐ family   ☐ team
☐ group   ☐ herd   ☐ crowd

Expanded nouns

Often, the thing, phenomenon or idea being described in a piece of writing is too complex to be covered by a single noun. In such cases the noun is 'expanded' by using a group of words[1] to function as a noun in the sentence. These groups of words can usually be replaced by the word 'something' or 'someone'.

Examples

☐ Students should know (something) *that a metre is longer than a yard.*

☐ (Something) *A mysterious plague among the owls at Johannesburg Zoo* has recently claimed 55 victims.

☐ One of the features of oxygen is (something) *that it is a colourless gas.*

☐ The main reason for this appears to be (something) *the fact that in many places there are few if any plants.*

☐ Adjectives

Definition

An adjective is a word which describes, qualifies or modifies a noun or pronoun.

Examples

The italicised words in the following paragraphs all function as adjectives.

In *these* days of *pocket* computers, *manual* aids to calculating such as the abacus seem *slow* and *dated*. Yet it is still used in *many* parts of the world, notably in China.

Many mechanical calculating machines have been produced since Pascal, the *French* mathematician, made his in 1642, but the *great* breakthrough did not come until the 1940s when the *first electronic* computer was made.

Bolitho, A R & P L Sandler
Study English for Science Longman 1989 56

---

1 A group or string of words is called a 'clause' if it contains a verb, and a 'phrase' if it does not contain a verb.

Biology is usually divided into *Plant* Biology or Botany, and *Animal* Biology or Zoology. However, *all* organisms, both plants and animals, are *similar* in *several important* ways. For example, *each* kind of organism has its own *particular* shape, or morphology. *Plant* Morphology and *Animal* Morphology are therefore the branches of Botany and Zoology in which we study the *external* shape of plants and of animals.

Pearson, Ian

English in Biological Science Oxford University Press 1978 1

Features of adjectives

Adjectives, as a class, have the following features which will help you to identify them:

(1) Adjectives answer the question 'what kind(s)?' or 'which one(s)?' about nouns

☐ what kind of aids? – *manual* aids
☐ which organisms? – *all* organisms

(2) Adjectives can tell us – amongst other things – about the quantity ('two'), size ('big'), shape ('round'), condition ('cracked'), age ('old'), colour ('red'), and substance ('china') of an object. Adjectives can also be evaluative as well as descriptive ('a *possible* explanation', 'a *doubtful* hypothesis', 'a *sound* theory').

(3) An adjective usually appears before the noun it is describing.

☐ *Plant* Biology   ☐ *external* shape   ☐ *important* ways

(4) An adjective may also appear alone (*ie* without being followed by a noun).

☐ the question is *difficult*   ☐ the shape is *anomalous*

(5) Adjectives do not have a singular or plural form.

(6) Adjectives can be formed by adding certain suffixes (word endings) to other words. The adjectival suffixes include:

☐ -ful (plenty + ful = plentiful)
☐ -al (logic + al = logical)
☐ -ive (act + ive = active)
☐ -able (depend + able = dependable)
☐ -y (scare + y = scary)
☐ -like (child + like = childlike)
☐ -ible (contempt + ible = contemptible)
☐ -ic (sulphur + ic = sulphuric)
☐ -ary (literature + ary = literary)

(7) Adjectives may appear in three forms, called the Degrees of Comparison:

Positive	Comparative	Superlative
good	better	best
simple	simpler	simplest
efficient	more efficient	most efficient

Practice session 2

Underline all the adjectives in the following paragraphs. Again, do not concern yourself so much with the meaning of the words, as with their form and function in the sentences.

(1) A hydraulic press is a device for securing mechanical advantage from the pressures of liquid columns. In its simplest form it consists of a wide cylinder and a narrow cylinder connected by a tube. Both cylinders are filled with water and fitted with a piston.

Allen, J B & H G Widdowson
English in Physical Science Oxford University Press 1979 52

(2) The input unit digests the program as it is received and passes it on to the store in the form of magnetic recordings. The store can supply information, or data, either to the arithmetic unit, the control unit or directly to the output unit, as it is needed. All the calculations are done in the arithmetic unit. The control unit, which is the 'nerve centre' of the computer, carries out the instructions which are received in a program. The arithmetic unit and the control unit each have small stores of information called registers. The output unit presents the results of each computer operation, usually as a printout.

Bolitho, AR & P L Sandler
Study English for Science Longman 1989 57

(3) Most species of Euglena possess plastids with chlorophyll inside (ie chloroplasts) and therefore they can photosynthesize. However, all the green species are unable to synthesize at least one organic substance that they need, and they must obtain these substances osmotrophically. They are therefore partly autotrophic and partly heterotrophic. The colourless species must obviously be fully heterotrophic.

Pearson, Ian
English in Biological Science Oxford University Press 1978 13

Check your answers against those at the end of this chapter before continuing.

Expanded adjectives   Sometimes more than one word is needed to provide a full and adequate description of a noun. In such cases, writers 'expand' the adjective by using a string of words.[2] This string functions as an adjective

---

2 A string of words functioning as an adjective is called an 'adjectival clause' if it contains a complete verb, and an 'adjectival phrase' if it does not.

in that it operates by describing, qualifying or modifying a noun, giving more specific and detailed information about that noun. Expanded adjectives, while they may contain a great number of words, still answer to the basic requirement of adjectives, namely that they answer the question 'which one(s)?' or 'what kind(s)?' about a noun.

Examples

☐ The solution *which is formed as a result of the process* is said to be neutral, since it does not turn blue litmus red, or turn red litmus blue.

(The clause in italics functions as an adjective answering the question 'which one?' about the noun 'solution'.)

☐ The white solid *formed in this action* is the salt ammonium chloride *which is formed by direct combination of the two gases.*

(The first clause in italics functions as an adjective defining the noun 'solid', while the second clause describes 'ammonium chloride'.)

☐ Movement, growth and reproduction are processes *which require a great deal of energy.*

☐ If the temperature of oxygen falls below -183° C, it changes from a colourless gas to a bluish liquid, *which is highly magnetic.*

☐ This can be shown experimentally by giving the same quantities of food to two different animals or groups of animals *that have the same overall body weight.* ... Such experiments are particularly important in the case of those animals *that are used by man as a source of meat.*

☐ The control unit, *which is the 'nerve centre' of the computer,* carries out the instructions *which are received in a program.*

Adjectival information is often introduced by one of the following connecting words:

☐ who   ☐ whom   ☐ which   ☐ that

Practice session 3

Underline the expanded adjectives in the following sentences. Circle the noun which is being described in each case.

(1) Vitamin B consists of twelve different chemicals, which are found in eggs, cheese, butter, wholemeal flour and vegetables. ... Vitamin D is the only vitamin which the body can make for itself, but it can only do this if there is sufficient sunlight. A lack of both sunlight and vitamin D can result in a disease called rickets, which causes bones to soften and be deformed.

Bates, Martin & Tony Dudley-Evans
Nucleus – General Science Longman 1990 72

continued

---

continued

(2) The method is similar to the one that was described on page 78, as the following experiment will indicate.

(3) Many substances which are bleached by moist chlorine are not bleached by exposure to ordinary oxygen.

(4) There are many adaptations that are shared by only a small number of species, and there are some that are found in just one or two species.

(5) Substances whose molecules are composed of atoms of other substances are known as compounds. Other substances have molecules which cannot be broken down into atoms of other substances, and these are called elements.

(6) Electrolysis was first investigated by Michael Faraday, who investigated the relation between the quantity of electricity passing through an electrolyte and the quantity of substance liberated by it.

Check your work against the answers at the end of this chapter before continuing.

Expanded adjectives deserve close attention in your reading. Very often they provide crucial defining information (as in (5) above), or they give explanatory material, examples or details which give you a more thorough understanding of a concept (as in (1) above, in which the term 'rickets' is explained).

☐ Verbs

Definition

A verb is a 'doing' word whose function is to express either an action or a state of being.

Examples

The italicised words in the following paragraphs are all verbs:

If we *put* some dilute sulphuric acid in a beaker and *dip* a plate of pure zinc in it, there *is* no visible action. If we *dip* a plate of copper into the acid, but without *allowing* it to touch the zinc plate, still no action *will be observed*. However, if we *allow* the zinc and copper plates to *touch* one another, at once bubbles of hydrogen *will rise* from the copper plate. Now *connect* wires to the terminals of a flash-lamp bulb and *connect* a zinc plate to one wire and a copper plate to the other. *Dip* the plates into dilute sulphuric acid, but *do* not *allow* them to *touch*. Bubbles of hydrogen *will rise* from the copper plate and the bulb *will glow, showing* that a current *is passing* through it.

Allen, J P B & H G Widdowson
English in Physical Science Oxford University Press 1979 81

How long an individual *survives depends* partly on chance and partly on whether it *has* any advantage over other individuals. For example, when food *is* short, the better hunter, or the faster eater, or the larger individual *may survive* when others *cannot*. Some individuals, in other words, *have* characteristics that *have* a survival value and we *say* that they *are* better adapted to their environment than the others. Obviously, the individuals which *are* better adapted *have* a higher chance of *living* long enough to *reproduce*. Less well-adapted individuals *are* more likely *to meet* death before *reaching* maturity.

Pearson, Ian
English in Biological Science Oxford University Press 1978 67

Features of verbs

(1) Verbs can describe an action

☐ touch ☐ connect ☐ glow
☐ reproduce ☐ survive

(2) Verbs can link ideas. (They are then called linking verbs.)

☐ there *is* no reaction ☐ individuals *are* better adapted

(3) Verbs can be complete – see points (4), (5), and (6) – or incomplete:

☐ larger individuals *may survive*
no reaction *occurred* } complete
some individuals *have adapted* for survival

☐ without *allowing* it *to touch* the plate
before *reaching* maturity } incomplete
individuals have a chance of *surviving*

(4) A complete verb has tense, ie it tells you about the time when the action occurred or when the state of affairs prevailed

☐ there *is* no visible reaction – present
☐ bubbles *will rise* from the plate – future
☐ the bulb *glowed* when the current *passed* through it – past

(5) A complete verb changes form to agree with its subject

☐ an individual *survives* (singular subject, singular form of verb)
many individuals *survive* (plural subject, plural form of verb)

☐ I *am* intelligent
you *are* intelligent } agreement of person
he / she / it *is* intelligent

(6) Verbs can be 'helping' words which combine with other verbs to make a complete verb. (These are called auxiliary verbs.)

☐ may ☐ will/shall ☐ should
☐ ought ☐ have ☐ was/were

(7) A verb can denote an action or state of being in one of four moods.
Indicative mood : the verb states a fact or asks a question

☐ there *is* no reaction
☐ *can* we *predict* the effect?

Imperative mood : the verb expresses a command

☐ *connect* wires to the terminals
☐ *dip* the plates into acid, but *do not allow* them to touch

Infinitive (the 'to ...' form) : the verb makes a statement in a
general way

☐ most individuals want *to survive*
☐ they live long enough *to reproduce*
☐ they are likely *to meet* death

Subjunctive mood : the verb expresses a wish, a condition, a purpose, or a doubt

☐ if I *were* you, I would ...
☐ he proposed that the plan *be shelved*
☐ take care lest the reaction *run* out of control
☐ she referred to the species as if it *were* already extinct

---

Practice session 4

Underline all the verbs in the following paragraphs.

(1) When a heavy body such as a stone or a piece of iron is suspended on a string and
totally immersed in water, there are three forces acting on it: the downward force
due to the weight of the object, the upthrust due to the displaced water, and the
tension of the supporting string.

Allen, J B & H G Widdowson
English in Physical Science Oxford University Press 1979 37

(2) Most species of Euglena possess plastids with chlorophyll inside (*ie* chloroplasts) and
therefore they can photosynthesize. However, all the green species are unable to
synthesize at least one organic substance that they need, and they must obtain these
substances osmotrophically. They are therefore partly autotrophic and partly heterotrophic. The colourless species must obviously be fully heterotrophic.

Pearson, Ian
English in Biological Science Oxford University Press 1978 13

(3) Let us consider an example of a process which illustrates the relationship between
acids, bases, and salts. If a solution of sodium hydroxide is poured into a beaker
containing hydrochloric acid and red litmus, the litmus will turn blue. This shows
that the acid has been neutralized and that sodium hydroxide is a base. As a result
of the neutralization of the acid by the base, sodium chloride is formed and may
be obtained as crystals by evaporation of the solution. Thus the result of adding
the base sodium hydroxide to hydrochloric acid is to form a salt, sodium chloride.

———— *continued* ————

continued

This substance forms as a result of the metal sodium replacing the hydrogen in the acid.

Allen, J B & H G Widdowson
English in Physical Science Oxford University Press 1979 23

(4) Animals ... must obtain complex organic substances by eating plants or other animals. The reason for this is that they lack chlorophyll. Among these 'other-feeders' or heterotrophs, we distinguish between 'solid-feeders' or phagotrophs, and 'liquid-feeders', or osmotrophs. Whereas phagotrophic organisms take in solid and often living food, osmotrophic ones absorb or suck up liquid food. This is usually from dead and rotting organisms.

Pearson, Ian
English in Biological Science Oxford University Press 1978 13

Check your answers against those at the end of this chapter before continuing to the next section.

☐ Adverbs

Definition

An adverb is a word that modifies or describes a verb, an adjective, another adverb, or a whole sentence.

Examples

The italicised words in the sentences below are all adverbs.

Most species of Euglena possess plastids with chlorophyll inside (*ie* chloroplasts) and therefore they can photosynthesize. However, all the green species are unable to synthesize at least one organic substance that they need, and they must obtain these substances *osmotrophically*. They are therefore *partly* autotrophic and *partly* heterotrophic. The colourless species must obviously be *fully* heterotrophic.

Pearson, Ian
English in Biological Science Oxford University Press 1978 13

How long an individual survives depends *partly* on chance and *partly* on whether it has any advantage over other individuals. For example, when food is short, the better hunter, or the faster eater, or the larger individual may survive when others cannot. Some individuals, in other words, have characteristics that have a survival value and we say that they are *better* adapted to their environment than the others. *Obviously,* the individuals which are *better* adapted have a higher chance of living *long enough* to reproduce. *Less well*-adapted individuals are *more likely* to meet death before reaching maturity.

Pearson, Ian
English in Biological Science Oxford University Press 1978 67

Features of adverbs

(1) Adverbs provide the answer to the questions 'how?', 'when?', 'where?' and 'why?'.

(2) Adverbs which answer the question 'how?' are called adverbs of manner.

☐ ... they must obtain these substances *osmotrophically*
☐ The weight can be raised *easily* ...
☐ ... the stimulus changed *gradually*.

(3) Adverbs which answer the question 'when?' are called adverbs of time.

☐ ... he arrived *yesterday*
☐ Suppose, *now*, that we want to find the relative density ...

(4) Adverbs which answer the question 'where?' are called adverbs of place.

☐ ... the tendency is to move *upwards*
☐ ... displacement occurs *sideways*

(5) Adverbs of degree provide information about the extent of the action of the verb (or the degree of meaning of the word or sentence which the adverb is modifying).

☐ How long an individual survives depends *partly* on chance and *partly* on whether it has any advantage over other individuals ...
☐ The experiment was *completely* successful.
☐ The colourless species must obviously be *fully* heterotrophic.

(6) Adverbs can be used for emphasis, expressing affirmation, doubt, or negation.

☐ *Obviously*, the individuals which are better adapted will ...
☐ *Certainly*, some oxidation occurs, ...
☐ *Probably*, the reason for this is ...
☐ ... the results are *not* applicable to all cases

(7) Most single-word adverbs end in -ly.

Practice session 5

Underline the adverbs in the following sentences. (Check whether the adverb you underline in each sentence provides the answer to the question 'how?' 'when?', 'where?' or 'why?')

(1) The wax in contact with the sides of the tube solidifies first.

(2) The temperature is lowered sufficiently for liquid air to be produced.

continued

—— *continued* ——

(3) Most biologists believe that the descendants of a group of reproducing organisms can become very different from their ancestors.

(4) Plants are characteristically autotrophic but animals are characteristically heterotrophic.

(5) Animals are directly or indirectly dependent upon plants.

(6) The relatively inefficient combustion of fuel and air in a car engine means that hydrocarbon fragments are left unburned.

(7) Carbon monoxide is a highly poisonous gas. Prolonged inhalation significantly reduces the oxygen-carrying capacity of the blood, and soon causes headaches, sickness, and – possibly – death.

(8) Parasites are never helpful, and are often harmful, to the host organism.

(9) Certain gases, such as ammonia, condense and re-evaporate quickly when the pressure on them varies.

(10) The temperature is controlled automatically by a thermostat switch.

Check your answers against those at the end of this chapter before continuing to the next section.

Expanded adverbs

If a single word is inadequate, adverbs can be expanded into strings of words.[3] These strings function as adverbs by answering the questions 'how?', 'when?', 'where?' and 'why?'.

Examples

☐ *When a heavy body such as a stone or a piece of iron is suspended on a string and totally immersed in water,* there are three forces acting on it: the downward force due to the weight of the object (W), the upthrust (w) due to the displaced water, and the tension (T) of the supporting string. *When the suspended object is in equilibrium,* $W = T + w$.

Allen, J P B & H G Widdowson
English in Physical Science Oxford University Press 1979 37

Before organisms die they usually produce offspring.

As the can cools down the steam inside condenses.

The italicised strings above are all expanded adverbs of time, answering the question *'when?'*.

☐ Animals must obtain complex organic substances by eating plants or other animals *because they lack chlorophyll.*

---

3 A string of words functioning as an adverb is called an 'adverbial clause' if it contains a complete verb, and an 'adverbial phrase' if it does not.

We can make general statements about plants and animals by comparing Ulothrix zonata with a rotifer, *because the former shows the typical plant characteristics and the latter the typical animal characteristics.*

The strings in italics above function as adverbs because they answer the question '*why?*'.

☐ We can show that air has weight *by means of the following experiment.*

The volume of a pyramid can be calculated *by a mathematical formula.*

The strings of words in italics above are functioning as expanded adverbs of manner, because they answer the question '*how?*'.

Expanded adverbs are often introduced by one of the following words:

after	although	as
because	before	if
since	though	until
unless	when	while
where	whenever	whether
now that	wherever	

Practice session 6

Underline the expanded adverbs in the following sentences. (NB There may be more than one expanded adverb in each sentence – make sure to underline them separately.)

(1) When paraffin wax is melted in a boiling tube and allowed to cool, the wax in contact with the sides of the tube solidifies first.

(2) The process continues until the temperature is lowered sufficiently for liquid air to be produced.

(3) If an attempt is made to increase the pressure of a saturated vapour by compression some of the vapour condenses.

(4) The cycle of evaporation and condensation continues as long as the refrigerator is switched on.

(5) Food should never be re-frozen after it has been thawed out as it can never be quick-frozen in a home refrigerator.

(6) At the start of the experiment, the total body weights of both groups of animals were measured.

(7) This similarity is difficult to explain unless we accept that all vertebrates have a common ancestor.

continued

---
continued

(8) If an object is placed on a table-top the pressure between the object and the table can be measured according to a prescribed formula.

(9) Certain gases, such as ammonia, condense and re-evaporate quickly when the pressure on them varies.

(10) The temperature is controlled automatically by a thermostat switch.

Check your answers against those at the end of this chapter.

Expanded adverbs are extremely important in sentences. Train yourself to notice them, because they often qualify or restrict the meaning of the sentence, by stating conditions and limitations.

☐ Summary

In this chapter we have been looking at individual word classes : nouns, verbs, adjectives and adverbs. Remember the definitions:

A *noun* is a word naming a person, place, thing, phenomenon, idea, quality, or condition.

An *adjective* is a word which describes, qualifies or modifies a noun or pronoun.

A *verb* is a 'doing' word whose function is to express either an action or a state of being.

An *adverb* is a word that modifies or describes a verb, an adjective, another adverb, or a whole sentence.

It is important to remember that the same word can fall into different word classes depending on how it functions. The word 'light', for example, can function as a noun ('Let there be light'), a verb ('It is better to light a candle than to curse the darkness'), an adjective ('A light load is easier to carry than a heavy one') or an adverb ('a light-blue coat').

In the next chapter, we will examine how words are put together to make statements of meaning – sentences.

☐ Answers

Practice session 1 – Nouns

(1) A hydraulic *press* is a *device* for securing mechanical *advantage* from the *pressures* of liquid *columns*. In its simplest *form* it consists of a wide *cylinder* and a narrow *cylinder* connected by a *tube*. Both *cylinders* are filled with *water* and fitted with a *piston*.

(2) Most *species* of *Euglena* possess *plastids* with *chlorophyll* inside (ie *chloroplasts*) and therefore they can photosynthesise. However, all the green *species* are unable to synthesise at least one organic *substance* that they need, and they must obtain these *substances* osmotrophically. They are therefore partly autotrophic and partly heterotrophic. The colourless *species* must obviously be fully heterotrophic.

(3) Let us consider an *example* of a *process* which illustrates the *relationship* between *acids, bases,* and *salts*. If a *solution* of *sodium hydroxide* is poured into a *beaker* containing *hydrochloric acid* and red *litmus*, the *litmus* will turn blue. This shows that the *acid* has been neutralised and that *sodium hydroxide* is a *base*. As a *result* of the *neutralisation* of the *acid* by the *base, sodium chloride* is formed and may be obtained as *crystals* by *evaporation* of the *solution*. Thus the *result* of adding the *base sodium hydroxide* to *hydrochloric acid* is to form a *salt, sodium chloride*. This *substance* forms as a *result* of the *metal sodium* replacing the *hydrogen* in the *acid*.

(4) *Animals* ... must obtain complex organic *substances* by eating *plants* or other *animals*. The *reason* for this is that they lack *chlorophyll*. Among these *'other-feeders'* or *heterotrophs,* we distinguish between *'solid-feeders'* or *phagotrophs,* and *'liquid-feeders',* or *osmotrophs*. Whereas phagotrophic *organisms* take in solid and often living *food,* osmotrophic *ones* absorb or suck up liquid *food*. This is usually from dead and rotting *organisms*.

Practice session 2 – Adjectives

(1) A *hydraulic* press is a device for securing *mechanical* advantage from the pressures of *liquid* columns. In its *simplest* form it consists of a *wide* cylinder and a *narrow* cylinder connected by a tube. *Both* cylinders are filled with water and fitted with a piston.

(2) The *input* unit digests the program as it is received and passes it on to the store in the form of *magnetic* recordings. The store can supply information, or data, either to the *arithmetic* unit, the *control* unit or directly to the *output* unit, as it is needed. *All* the calculations are done in the *arithmetic* unit. The *control* unit, which is the *'nerve* centre' of the computer, carries out the instructions which are received in a program. The *arithmetic* unit and the *control* unit each have

———— *continued* ————

— *continued* —

small stores of information called registers. The *output* unit presents the results of *each computer* operation, usually as a printout.

(3) *Most* species of Euglena possess plastids with chlorophyll inside (ie chloroplasts) and therefore they can photosynthesise. However, *all* the *green* species are unable to synthesise at least *one organic* substance that they need, and they must obtain *these* substances osmotrophically. They are therefore partly *autotrophic* and partly *heterotrophic*. The *colourless* species must obviously be fully *heterotrophic*.

Practice session 3 – Expanded Adjectives

(1) Vitamin B consists of twelve different chemicals, *which are found in eggs, cheese, butter, wholemeal flour and vegetables.* ... Vitamin D is the only vitamin *which the body can make for itself,* but it can only do this if there is sufficient sunlight. A lack of both sunlight and vitamin D can result in a disease called rickets, *which causes bones to soften and be deformed.*

(2) The method is similar to the one *that was described on page 78,* as the following experiment will indicate.

(3) Many substances *which are bleached by moist chlorine* are not bleached by exposure to ordinary oxygen.

(4) There are many adaptations *that are shared by only a small number of species,* and there are some *that are found in just one or two species.*

(5) Substances *whose molecules are composed of atoms of other substances* are known as compounds. Other substances have molecules *which cannot be broken down into atoms of other substances,* and these are called elements.

(6) Electrolysis was first investigated by Michael Faraday, *who investigated the relation between the quantity of electricity passing through an electrolyte and the quantity of substance liberated by it.*

Practice session 4 – Verbs

(1) When a heavy body such as a stone or a piece of iron *is suspended* on a string and totally *immersed* in water, there *are* three forces *acting* on it: the downward force due to the weight of the object, the upthrust due to the displaced water, and the tension of the supporting string.

(2) Most species of Euglena *possess* plastids with chlorophyll inside (*ie* chloroplasts) and therefore they can *photosynthesise*. However, all the green species *are unable to synthesise* at least one organic substance that they *need,* and they *must obtain* these substances osmotrophically. They *are* therefore partly autotrophic and partly heterotrophic. The colourless species *must* obviously *be* fully heterotrophic.

— *continued* —

— *continued* —

(3) *Let* us *consider* an example of a process which *illustrates* the relationship between acids, bases, and salts. If a solution of sodium hydroxide *is poured* into a beaker *containing* hydrochloric acid and red litmus, the litmus *will turn* blue. This *shows* that the acid *has been neutralised* and that sodium hydroxide *is* a base. As a result of the neutralisation of the acid by the base, sodium chloride *is formed* and *may be obtained* as crystals by evaporation of the solution. Thus the result of *adding* the base sodium hydroxide to hydrochloric acid *is* to *form* a salt, sodium chloride. This substance *forms* as a result of the metal sodium *replacing* the hydrogen in the acid.

(4) Animals ... *must obtain* complex organic substances by *eating* plants or other animals. The reason for this *is* that they *lack* chlorophyll. Among these 'other-feeders' or heterotrophs, we *distinguish* between 'solid-feeders' or phagotrophs, and 'liquid-feeders', or osmotrophs. Whereas phagotrophic organisms *take in* solid and often living food, osmotrophic ones *absorb* or *suck up* liquid food. This *is* usually from dead and rotting organisms.

Practice session 5 – Adverbs

(1) The wax in contact with the sides of the tube solidifies *first*.

(2) The temperature is lowered *sufficiently* for liquid air to be produced.

(3) Most biologists believe that the descendants of a group of reproducing organisms can become *very* different from their ancestors.

(4) Plants are *characteristically* autotrophic but animals are *characteristically* heterotrophic.

(5) Animals are *directly* or *indirectly* dependent upon plants.

(6) The *relatively* inefficient combustion of fuel and air in a car engine means that hydrocarbon fragments are left unburned.

(7) Carbon monoxide is a *highly* poisonous gas. Prolonged inhalation *significantly* reduces the oxygen-carrying capacity of the blood, and *soon* causes headaches, sickness, and – *possibly* – death.

(8) Parasites are *never* helpful, and are *often* harmful, to the host organism.

(9) Certain gases, such as ammonia, condense and re-evaporate *quickly* when the pressure on them varies.

(10) The temperature is controlled *automatically* by a thermostat switch.

Practice session 6 – Expanded Adverbs

(1) *When paraffin wax is melted in a boiling tube and allowed to cool,* the wax in contact with the sides of the tube solidifies first. (time)

(2) The process continues *until the temperature is lowered sufficiently* for liquid air to be produced. (time)

(3) *If an attempt is made to increase the pressure of a saturated vapour* (reason) *by compression* (manner) some of the vapour condenses.

(4) The cycle of evaporation and condensation continues *as long as the refrigerator is switched on.* (time)

(5) Food should never be re-frozen *after it has been thawed out* (time) *as it can never be quick-frozen* (reason) *in a home refrigerator.* (place)

(6) *At the start of the experiment* (time), the total body weights of both groups of animals were measured.

(7) This similarity is difficult to explain *unless we accept that all vertebrates have a common ancestor.*

(8) *If an object is placed on a table-top* the pressure between the object and the table can be measured *according to a prescribed formula.* (manner)

(9) Certain gases, such as ammonia, condense and re-evaporate quickly *when the pressure on them varies.* (time and manner)

(10) The temperature is controlled automatically *by a thermostat switch.* (manner)

2

Sentences

In Chapter 1, we looked at words. Words on their own, however, are not very expressive of concepts, ideas, theories, or events. That is why we speak and write in sentences – if words are units of meaning, sentences are units of thought.

☐ Sentences

Definition

> A sentence is an organised sequence of words that presents at least one complete thought.

Examples

The groups of words below are *not* sentences because they are not organised into a meaningful sequence.

☐ a can liquid be into changed a gas
☐ about ago millions evolved of 35 Mesohippus years
☐ only on it digits had forelimb three its
☐ string iron suspended a on string piece a of is

The sequences below are *not* sentences because they express a fragment or piece of an idea, instead of a complete idea.

☐ Although it had only three digits on its forelimb ...
☐ When a piece of iron is suspended on a string ...
☐ That a liquid can be changed into a gas ...
☐ Mesohippus, which evolved about 35 millions of years ago ...

The groups of words below are sentences; they are organised into a meaningful sequence, and they each present at least one complete thought.

☐ A liquid can be changed into a gas.
☐ Mesohippus evolved about 35 millions of years ago.

☐ Although it had only three digits on its forelimb, a fifth was represented by a splint bone.

☐ When a piece of iron is suspended on a string, two forces are acting on it.

Practice session 1

Circle the numbers of the sequences below which are sentences.

(1) By the middle of the Miocene, soft grasses replaced the broad-leaved plants.

(2) The downward force due to the weight of the object and the tension of the supporting string.

(3) Eocene climate the those the and vegetation were similar to of the.

(4) Totally immersed in water.

(5) There are three forces acting on it.

(6) Proteins, which are complex chemical subtances which give us energy.

(7) This process, called digitising, is carried out by a reading device called a geameter.

(8) Map stored is a long co-ordinates a as list of.

(9) Early 1960s: transistors – computers much smaller and easier to repair and maintain; first data processing; use in communications.

(10) But ultrasonic sound waves, with frequencies up to 10 million Hz, can also be produced by special apparatus.

Check your answers against those at the end of this chapter before continuing to the next section.

Features of sentences

(1) A sentence accomplishes at least two things:

☐ it names something (a thing, idea, person, place etc)
☐ it says something about what it has just named (for example, what the person, idea, or thing does, what it is, or what has been done to it)

(2) A sentence thus has two main parts :

☐ the subject (the person, thing, idea etc which is named)
☐ the predicate (the part of the sentence which says something about the subject)

☐ Subjects

Definition

> The subject of a sentence is the person, place, thing, idea, quality or condition that acts, or that is described or identified in the sentence.

Examples

Mesohippus evolved about 35 millions of years ago. *It* had only three digits on its forelimb.

Most species of Euglena possess plastids with chlorophyll inside.

A solution of sodium hydroxide is poured into a beaker containing hydrochloric acid.

The climate and the vegetation were still similar to those of the Eocene.

The normal range of human hearing is between the frequencies of 20 and 20 000 Hz.

The reason for this is that they lack chlorophyll.

Features of subjects

(1) You can identify the subject of the sentence by asking 'who' or 'what' is undergoing the action or being described by the sentence.

☐ Phagotrophic organisms take in solid and often living food.
Q: Who or what takes in solid food?
A: Phagotrophic organisms = the subject

☐ Above the electromagnet, and close to the head of the striker rod, is a gong.
Q: Who or what is above the electromagnet etc?
A: a gong = the subject

(2) The subject of the sentence always has a noun or pronoun as its core. (See Chapter 1's definition of nouns.)

(3) A subject may be **simple**, in which case it consists of a single word.

☐ *Positions* are given in degrees.
☐ *Compounds* are combinations of two or more elements.

(4) A **compound** subject consists of two or more simple subjects joined by a conjunction (joining or connecting word).

☐ *Plants and animals* use the air in the same way.
☐ *The core, the mantle, and the crust* form three concentric spheres or layers.

(5) The **complete** subject consists of the simple subject plus any words modifying, describing, or expanding it.

☐ *Substances which are attracted by a magnet* are known as magnetic substances.

☐ *The different paths traced out by the needle* can be represented by a diagram.

(6) A sentence may appear to have no subject.

☐ Place a magnetised needle in a cork.

☐ Calculate the area of the circle.

In these cases the pronoun 'you' is understood to be the subject.

Practice session 2

Underline the complete subjects of the sentences in the paragraph below.

An electric bell operates by means of an electromagnet. This consists of two cylinders of soft iron fixed one above the other to a soft iron bar. Around these cylinders is wound a length of copper wire, the direction of winding being reversed as the wire passes from one cylinder to another. One end of it passes from the free end of the upper cylinder and is connected to a battery terminal. The other end passes down from the top of the lower cylinder and is connected to the fixed end of a steel spring which is situated below and to the left of the electromagnet. To the right side of this spring is attached a metal rod, the head of which acts as a hammer, or striker. The spring passes up the left side of the striker rod and then bends outwards to touch a screw, or key, which is connected to the other terminal of the battery by means of copper wire. On the other side of the striker rod, just opposite the free ends of the soft iron cylinders, is fixed a piece of soft iron which is called the armature. Above the electromagnet, and close to the head of the striker rod, is a gong.

Allen, J P B & H G Widdowson
English in Physical Science Oxford University Press 1979 82

Check your answers against those at the end of the chapter before continuing to the next section.

☐ Predicates

Definition

> The predicate of a sentence is the part of the sentence which expresses the action or state of being of the subject.

Examples

The italicised parts of the sentences below are the predicates of the sentences.

☐ Certain gases *condense and re-evaporate quickly when the pressure on them varies.*

☐ Plants *use carbon dioxide and water to produce carbohydrates.*

☐ A block of metal of mass 10 kg resting on a flat surface such as a table *exerts a downward force of about 98 newton.*

☐ A human being under normal conditions *emits about 63 g of water vapour per hour.*

Features of predicates

(1) The predicate of a sentence states what the subject of the sentence does, what it is, or what has been done to it.

(2) The predicate of a sentence always contains at least one verb.

(3) The predicate of a sentence may contain an object (see later in this chapter).

Practice session 3

Underline the predicates in the following sentences. (Hint: Eliminate the subject and all the adjectival information about the subject. What remains is the predicate – the statement about the subject.)

(1) A bimetallic strip is used in central heating systems.

(2) The hot gas passes into a condenser where it is cooled and condensed to liquid ammonia.

(3) Heat and light energy are propagated by transverse waves. The most common example of such waves is found in water.

(4) Citric acid, which is found in lemons and oranges and other citrus fruits, and acetic acid, which is found in vinegar, are organic acids.

(5) Some liquids which act as conductors of electricity decompose when an electric current is passed through them.

(6) Birds which eat worms which have eaten leaves which have been sprayed with insecticide which contains aldrin, which is a substance toxic to animals, are likely to die or become ill.

Check your answers against those at the end of this chapter before continuing to the next section.

☐ Objects

Definition

The object of a sentence is the person or thing directly affected by the verb; it is the direct receiver of the action of the verb.

Examples

The italicised words in the sentences below are the objects of the sentences.

- ☐ By the middle of the Miocene, soft grasses replaced *the broad-leaved plants.*
- ☐ Most species of Euglena possess *plastids with chlorophyll inside.*
- ☐ Phagotrophic organisms take in *solid and often living food.*

Features of objects

(1) You can identify the object of the sentence by asking 'who?' or 'what?' after the verb.

  ☐ Plants and animals use the air in the same way.
  Q: Who or what do plants and animals use?
  A: the air = the object

(2) The object of a sentence always has a noun or pronoun as its core. (See the definition of nouns in Chapter 1.)

(3) An object is not an essential part of a sentence. Many sentences do not contain objects.

☐ Summary

In this chapter, we have been looking at the main constituents of a sentence, and how they function together. A simple sentence can be depicted in the following way:

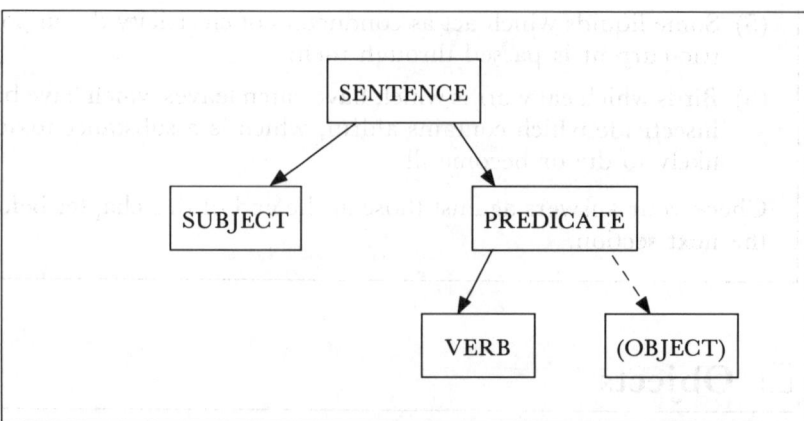

In other words, a sentence has to have a subject and a predicate, which consists of a compulsory verb and an optional object. This is the basic skeleton of a sentence – many more ideas may be added or inserted,

as you will see in the next chapter. At this point, however, it is important that you are familiar with the basic constituents and their definitions, which are repeated below.

> A sentence is an organised sequence of words that presents at
> least one complete thought.

> The subject of a sentence is the person, place, thing, idea, qual-
> ity or condition that acts, or that is described or identified in
> the sentence.

> The predicate of a sentence is the part of the sentence which
> expresses the action or state of being of the subject.

> The object of a sentence is the person or thing directly affected
> by the verb; it is the direct receiver of the action of the verb.

☐ Answers

Practice session 1 – Sentences

You should have circled the numbers of the following sequences; they are the only
sentences. The other sequences either do not present complete ideas, or the words
are not in a meaningful sequence.

(1) By the middle of the Miocene, soft grasses replaced the broad-leaved plants.

(5) There are three forces acting on it.

(7) This process, called digitising, is carried out by a reading device called a
geameter.

(10) But ultrasonic sound waves, with frequencies up to 10 million Hz, can also
be produced by special apparatus.

Practice session 2 – Subjects

An electric bell operates by means of an electromagnet. *This* consists of two cylinders
of soft iron fixed one above the other to a soft iron bar. Around these cylinders
is wound *a length of copper wire*. *The direction of winding* is reversed as the wire passes
from one cylinder to another. *One end of it* passes from the free end of the upper
cylinder and is connected to a battery terminal. *The other end* passes down from
the top of the lower cylinder and is connected to the fixed end of a steel spring

continued

---

continued

which is situated below and to the left of the electromagnet. To the right side of this spring is attached *a metal rod,* the head of which acts as a hammer, or striker. *The spring* passes up the left side of the striker rod and then bends outwards to touch a screw, or key, which is connected to the other terminal of the battery by means of copper wire. On the other side of the striker rod, just opposite the free ends of the soft iron cylinders, is fixed *a piece of soft iron* which is called the armature. Above the electromagnet, and close to the head of the striker rod, is *a gong.*

Practice session 3 – Predicates

(1) A bimetallic strip *is used in central heating systems.*

(2) The hot gas *passes into a condenser where it is cooled and condensed to liquid ammonia.*

(3) Heat and light energy *are propagated by transverse waves.* The commonest example of such waves *is found in water.*

(4) Citric acid, which is found in lemons and oranges and other citrus fruits, and acetic acid, which is found in vinegar, *are organic acids.*

(5) Some liquids which act as conductors of electricity *decompose when an electric current is passed through them.*

(6) Birds which eat worms which have eaten leaves which have been sprayed with insecticide which contains aldrin, which is a substance toxic to animals, *are likely to die or become ill.*

3

Ideas in Sentences

Thus far, we have been assuming that a sentence primarily expresses one main idea. This is true for simple writing, but the kind of texts you are involved in reading often express subtle and complex concepts and the sentences usually express a number of ideas. To the initial, basic, main idea of a sentence can be added –

☐ another one main idea or more
☐ adjectival information describing the nouns in the sentence
☐ adverbial information describing the verbs in the sentence, or qualifying the whole sentence.

Long, complicated sentences can be broken down again so that each added piece of information or idea is expressed in a simple sentence.

Examples

She went to breakfast and then stood smoking a cigarette at her desk, looking down at her books and the Study Guides that she had read so many times; and she wondered, rather desperately, whether she would ever achieve her dream of becoming a scientist – a dream which was beginning to seem increasingly foolish.

☐ She went to breakfast.
☐ She stood at her desk.
☐ She was smoking a cigarette.
☐ She looked down at her books.
☐ She looked down at her Study Guides.
☐ She had read them many times.
☐ She wondered.
☐ Her wondering was rather desperate.
☐ She had a dream of being a scientist.
☐ The dream was begining to seem increasingly foolish.

Birds which eat worms which have eaten leaves which have been sprayed with insecticide which contains aldrin, which is a substance toxic to animals, are likely to die or become ill.

☐ Birds are likely to die.
☐ Birds are likely to become ill.
☐ Birds eat worms.

☐ Worms eat leaves.
☐ Leaves have been sprayed with insecticide.
☐ Insecticide contains aldrin.
☐ Aldrin is a substance toxic to animals.

Breaking down a sentence into its individual ideas is not really necessary for sentences like the the two examples above — where the meaning is fairly clear. However, sentences in scientific texts are often very long and confusing, and the reader can easily get lost in the middle of them without ever being able to find the way out. If you find that you have reached the end of a sentence without understanding what it was trying to tell you, it is useful to go back and try to break it up into its constituents. In this way, you can distinguish the main idea(s) from all the supporting or qualifying detail, and improve your understanding of what the writer is trying to convey.

☐ Compound Sentences

Definition

A compound sentence is a sentence in which two or more main ideas are linked together.

Examples

☐ Idea 1 : By the middle of the Miocene the climate was drier.
  +
  Idea 2 : Soft grasses replaced the broad-leaved plants.
  =   By the middle of the Miocene the climate was drier and soft grasses replaced the broad-leaved plants.

☐ Idea 1 : Phagotrophic organisms take in solid and often living food.
  +
  Idea 2 : Osmostrophic organisms absorb or suck up liquid food.
  =   Whereas phagotrophic organisms take in solid and often living food, osmostrophic ones absorb or suck up liquid food.

☐ Idea 1 : Vitamin D is the only vitamin which the body can make for itself.
  +
  Idea 2 : The body can only do this if there is sufficient sunlight.
  =   Vitamin D is the only vitamin which the body can make for itself but it can only do this if there is sufficient sunlight.

Features of compound sentences

(1) Ideas can be linked together in compound sentences using linking words (also called conjunctions).

(2) Ideas are combined in different relationships, depending on the type of linking word (conjunction) used.

(3) In compound sentences, the linked main ideas have equal weight and are equally important.

(4) The ideas in compound sentences can simply be added to one another, using an *additive* linking word:

and	besides	moreover	furthermore

☐ We tried to make this book useful, *and* we did our best to make it interesting as well.

(5) The ideas in compound sentences can be set in opposition to one another, by using a *contrastive* linking word to show a contrast between the two ideas:

but	however	still	yet	nevertheless	otherwise

☐ We tried to make this book useful, *but* it is difficult to meet the needs of every possible student.

(6) The ideas in a compound sentence can be linked in a causal chain (one idea giving the cause or reason, and the other giving the result or effect) by using *cause and effect* linking words:

therefore	hence	consequently	thus	so

☐ We tried to make this book useful, *therefore* we covered a comprehensive range of language skills.

(7) The ideas in a compound sentence can be put into a chronological sequence by using *time* linking words:

afterward	earlier	next	then	later

☐ We first decided that the book should be useful, *thereafter* we agreed to try to make it interesting as well.

(8) Ideas can also be joined without using linking words, but using *punctuation* (often a semi-colon) instead, to mark the break between the ideas.

☐ We tried to make the book useful; we also tried to make it interesting.

Practice session 1

Read the following compound sentences carefully. Then write out each individual idea as a separate sentence. The first one has been done for you as an example.

(1) A heavy body , such as a large rock, can be raised easily when it is under water, but seems to be much heavier as soon as it comes out into the air.

   Idea 1: A heavy body, such as a large rock, can be raised easily when it is under water.

   Linking word: but

   Idea 2: It seems to be much heavier as soon as it comes out into the air.

(2) The sides of a cardboard carton of milk will often curve outwards, and this is caused by the sideways force exerted by the milk.

   Idea 1: ....

   ....

   Linking word: ....

   Idea 2: ....

   ....

(3) Dip the plates into dilute sulphuric acid but do not allow them to touch.

   Idea 1: ....

   ....

   Linking word: ....

   Idea 2: ....

   ....

(4) Carbon is present in the atmosphere only in small amounts, thus it has to be used again and again.

   Idea 1: ....

   ....

   Linking word: ....

   Idea 2: ....

   ....

———— *continued* ————

─── *continued* ───

(5) The level of the water surface in the measuring jar is read, then the solid is lowered into the vessel until it is completely covered by the water.

Idea 1: ....

....

Linking word: ....

Idea 2: ....

....

Linking word: ....

Idea 3 : ....

....

Check your answers against those at the end of this chapter before continuing to the next section.

☐ Complex Sentences

Additional information can be incorporated into sentences in other ways as well.

☐ The nouns can be qualified by adding descriptive information in the form of expanded adjectives.

☐ The verbs can be qualified by expanded adverbs, which can also be used to qualify the whole sentence.

Sentences containing additional phrases and or clauses of this kind are called complex sentences.

Definition

Complex sentences are sentences which contain at least one main idea, with a number of qualifying phrases or clauses.

Examples

☐ When a liquid presses against the container in which it rests, we say that it exerts a force.

Idea 1: A liquid presses against a container.
(Used as an expanded adverb qualifying the sentence.)

Idea 2: The liquid rests in the container.
(Used as an expanded adjective qualifying 'container')

Idea 3: We   say   something.
(subject)   (verb)   (object)

Idea 4: The liquid exerts a force.
(Used as an expanded noun, the object of the sentence.)

☐ When the zinc dissolves, each zinc ion leaves two electrons behind on the zinc plate, which consequently becomes negatively charged.

Idea 1: The zinc dissolves.
(Used as an expanded adverb qualifying the whole sentence.)

Idea 2: Each zinc ion leaves two electrons.
(subject)   (verb)   (object)

Idea 3: ...behind on the zinc plate
(an expanded adverb, answering the question 'where?' about the verb 'leaves')

Idea 4: The zinc plate becomes negatively charged in consequence.
(Used as an expanded adjective.)

Features of complex sentences

(1) Complex sentences contain at least one main idea, as well as any number of subsidiary ideas which add information or detail about any of the constituents of the main idea.

(2) The subsidiary information is often introduced by connecting words.

(3) Adjectival information may be introduced by:

who	whom	which	that

(4) Adverbial information may be introduced by:

after	although	as
because	before	if
since	though	until
unless	when	while
where	whenever	whether
now that	wherever	

(5) The subsidiary information may be in the form of a phrase (a string of words without a complete verb), or a clause (a string of words containing a complete verb).

Practice session 2

Write each separate piece of information in the following sentences as a separate sentence or phrase (as done in the examples above). You do not need to identify whether the ideas are used as adjectives, adverbs, etc. Merely write out the individual ideas.

(1) Some liquids which act as conductors of electricity decompose when an electric current is passed through them.

Idea 1: ....

....

Idea 2: ....

....

Idea 3: ....

....

(2) If a bar magnet is placed in iron filings, most of the filings will stick to the ends of the magnet.

Idea 1: ....

....

Idea 2: ....

....

Idea 3: ....

....

(3) A magnet sets in a definite direction when freely suspended.

Idea 1: ....

....

Idea 2: ....

....

Idea 3: ....

....

continued

─── *continued* ───

(4) If the copper and zinc plates touch one another or are connected by a conductor, excess electrons flow from the zinc plate to the copper plate, which is at a higher potential.

Idea 1: ....

....

Idea 2: ....

....

Idea 3: ....

....

Idea 4: ....

....

(5) If the other surface of the object measures 100cm² (0.01m²) and if this makes contact with the table-top, the pressure between the object and the table will be 200 N / 100 cm², which equals 20 000 N/m².

Idea 1: ....

....

Idea 2: ....

....

Idea 3: ....

....

Idea 4: ....

....

Check your answers against those at the end of this chapter.

☐ Summary

This chapter has focussed on the fact that a single sentence can present you with a great deal of information. Compound and complex sentences were defined and explained.

A compound sentence is a sentence in which two or more main ideas are linked together.

Complex sentences are sentences which contain at least one main idea, with a number of qualifying phrases or clauses.

It can often be useful to break sentences down into their constituent parts, so that you are sure that you arrive at the end of the sentence understanding all the 'bits' of information, as well as how they fit together, and how important each 'bit' is. Stating each 'bit' as an independent sequence can often help to make sense of a long and contorted sentence.

Another approach is to see a sentence as a track, with binary decision points at certain nodes, at which points the sentence can branch off to include additional information. An example of such a track could be the following:

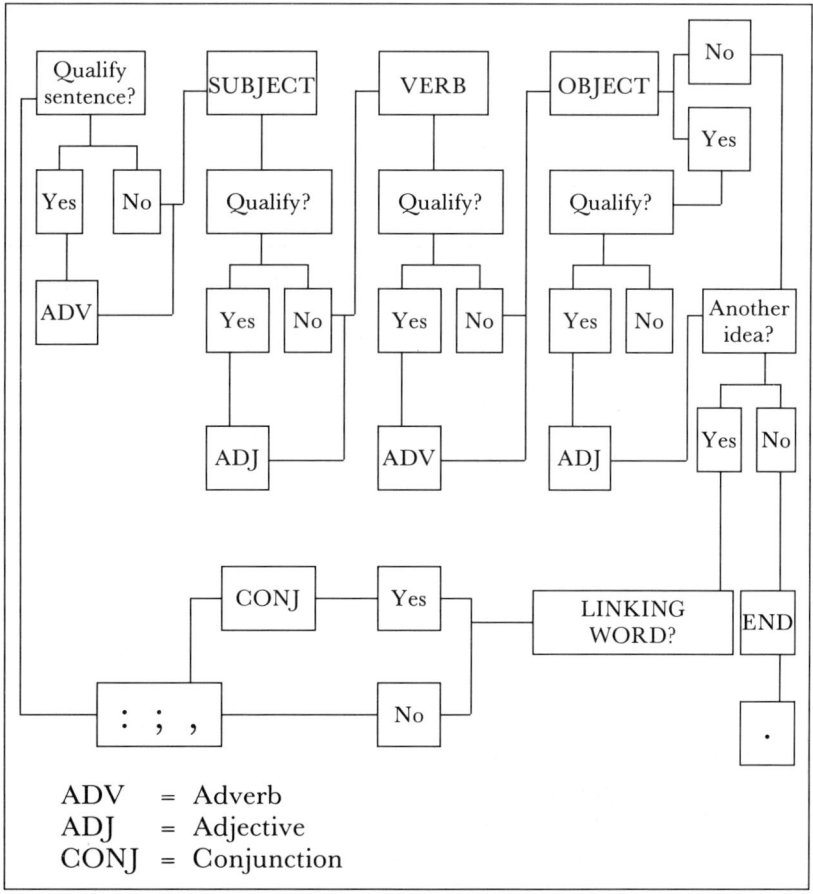

ADV   = Adverb
ADJ   = Adjective
CONJ = Conjunction

☐ **Answers**

Practice session 1 – Compound Sentences

(2) Idea 1: The sides of a cardboard carton of milk will often curve outwards.
Linking word: and
Idea 2: This is caused by the sideways force exerted by the milk.

(3) Idea 1: Dip the plates into dilute sulphuric acid.
Linking word: but
Idea 2: Do not allow them to touch.

(4) Idea 1: Carbon is present in the atmosphere only in small amounts.
Linking word: thus
Idea 2: It has to be used again and again.

(5) Idea 1: The level of the water surface in the measuring jar is read.
Linking word: then
Idea 2: The solid is lowered into the vessel.
Linking word: until
Idea 3: It is completely covered by the water.

Practice session 2 – Complex Sentences

(1) Idea 1: Some liquids decompose.
subject + verb = main idea
Idea 2: Liquids act as conductors of electricity.
(Used as an expanded adjective.)
Idea 3: An electric current is passed through them.
(Used as an expanded adverb.)

(2) Idea 1: A bar magnet is placed in iron filings.
(Used as an expanded adverb.)
Idea 2: Most of the filings will stick to the ends of the magnet.
(subject) (verb) (expanded adverb)

(3) Idea 1: A magnet sets.
subject + verb = main idea
Idea 2: ...in a definite direction ...
(expanded adverb)
Idea 3: It is freely suspended.
(Used as an expanded adverb.)

continued

---

continued

(4) Idea 1: The copper and zinc plates touch one another.
      (Used as an expanded adverb.)
    Idea 2: The copper and zinc plates are connected by a conductor.
      (Used as an expanded adverb.)
    Idea 3: Excess electrons flow from the zinc plate to the copper plate.
      (Main idea)
    Idea 4: The copper plate is at a higher potential.
      (Used as an expanded adjective.)

(5) Idea 1: The other surface of the object measures 100 cm² (0.01 m²).
      (Used as an expanded adverb.)
    Idea 2: This makes contact with the table-top.
      (Used as an expanded adverb.)
    Idea 3: The pressure between the object and the table will be 200 N/100 cm².
      (Main idea)
    Idea 4: This equals 20 000 N/m².
      (Used as an expanded adjective.)

4

Paragraphs

The previous chapters dealt with words and sentences, and the ideas expressed in these. However, you are seldom involved in reading single words or individual sentences. In studying the sciences at tertiary level, you are dealing with great masses of text, consisting of vast numbers of sentences. These masses of text are not just a formless bulk, however. On the page, they are visibly divided into smaller units, called paragraphs. It is these units with which we will be dealing in this chapter.

☐ Paragraphs

Definition

> A paragraph is a collection of two or more sentences dealing with a single idea, concept, or stage in the development of an argument.

Features and functions of paragraphs

(1) Paragraphs usually contain more than one sentence.

(2) Paragraphs tend to have a single main idea or central concern.

(3) *Introductory paragraphs* appear at the beginning of a chapter or section and may indicate:

    ☐ the central concern of the chapter or section
    ☐ the parameters of the subject to be dealt with
    ☐ how the subject is to be approached

Example

In this chapter, we discuss and give example solutions to two knowledge engineering issues that remain central to the successful development of intelligent tutoring systems: 1. representing domain knowledge, eg concepts and processes of the domain, and 2. representing teaching knowledge, eg instructional and discourse strategies to teach domain knowledge. We also address issues related to up-scaling existing intelligent tutoring technology to practical levels to bring tutoring systems into the real world.

Self, John (ed)
Artificial Intelligence and Human Learning Chapman and Hall 1988 3

Introductory paragraphs have an important function in that they prepare you to read what is to follow. In a way, they define the parameters of a subject, which the chapter or section will proceed to explain. A good reader will thus respond to an introductory paragraph with a state of mental preparedness. Often this will take the form of questions to oneself, which the text should then answer as one reads further.

(4) *Expository paragraphs* form the body of most texts. They present, explore, and discuss new information.

Example

The properties of a polymer depend on its molecular structure. Melting point, chemical reactivity, resistance to solvents, mechanical strength, and flexibility vary according to the type, number, and arrangement of monomer units. Polymers in which monomer units are arranged in nearly parallel chains form fibers which may be very strong and flexible. If the chains are tangled, the polymer may form as a strong film or bulk solid.

Masterton, W L *et al*
Chemistry Principles Saunders 1981 617

Expository paragraphs, thus, have the important function of presenting you with the 'meat', or major substance of the text. It is in the expository paragraphs that you will find the bulk of the information which you need to understand.

(5) *Transitional paragraphs* are usually brief paragraphs within the body of a text, whose function is to link the material together into a coherent whole, and to provide a bridge between material which has just been read and material that is about to be read.

Example

In the past two sections, we have seen how the formulas of molecular substances can be determined by experiment. We now move on to a study of how formulas of ionic compounds can also be found in the laboratory. It is also possible to predict 'on paper' the formulas of many simple ionic compounds.

Masterton, W L *et al*
Chemistry Principles Saunders 1981 55 (Slightly adapted)

The *function* of transitional paragraphs is to mark off distinct stages in the development of a text. Their function combines that of concluding paragraphs and introductory paragraphs in that they sum up what has been said, and prepare the ground for new material. If you fail to read transitional paragraphs closely, you might not understand how material fits together, or how the ideas in a text are connected.

(6) *Concluding paragraphs* appear at the end of a text or section of a text. Their function is to restate or summarise the main gist of the preceding text. Concluding paragraphs may list the central ideas of the text, draw conclusions from these ideas, or offer speculation as to the implications and possibilities of these ideas.

Example

In general, the knowledge that enables us to build intelligent tutors is not yet fully understood. Further research into each domain and tutoring knowledge is required to make further advances in this area. In this chapter we have suggested some ways that knowledge representations and control structures have been built to encode complete knowledge into tutoring systems. In addition, we suggest that sophisticated AI techniques, excruciating attention to detail, and intuitions about teaching and learning must continue to be researched and incorporated into tutors if we are to make progress in building more sophisticated machine tutors.

Self, John (ed)
Artificial Intelligence and Human Learning Chapman and Hall 1988 27

The *function* of concluding paragraphs is partly to provide you, the reader, with a checklist to control your own understanding of the preceding material. If a concluding paragraph says, as the above example does, that the chapter has dealt with the ways in which control structures have been built, and you have only a faint recollection and understanding of this, then obviously you need to re-read the relevant parts of the chapter. The concluding paragraph is thus useful in that it reflects what you should have read and come to understand in the course of dealing with the text.

☐ Main Ideas

We stated above that every paragraph has a single main idea or central concern. The sentences in a paragraph are grouped together precisely because they all deal with that single main idea. In many paragraphs, the main idea is expressed in a *topic sentence* – a sentence in the paragraph which explicitly states and encapsulates the main idea of that paragraph. The topic sentence is always the most *general* sentence in the paragraph.

Train yourself to identify the topic sentences in the paragraphs you read – they will serve as a condensed version of the paragraph. If a paragraph does not have an identifiable topic sentence, construct one by asking yourself: 'What is this paragraph about? What is it saying? What is the general idea?' (In constructing your own topic sentence, remember that a sentence must have a verb. Identifying the main idea of a paragraph as 'soap', for example, is not very useful. What is the paragraph saying *about* soap? Always constructing a topic sentence with a verb is one way to ensure that you have captured the main idea as fully as possible – as in 'Soap is a mixture of sodium salts of long-chain fatty acids', for example.)

Stopping at the end of each paragraph you read to summarise it in a single sentence is a good way of making texts more easily digestible, because you are breaking them down into more manageable units.

Practice session 1

Underline the most general statement in each of the groups of statements below. Check your answers against those at the end of this chapter before continuing to the next practice session.

(1) ☐ One system is the skeleton, which serves to support the body and protect the internal organs.
☐ The digestive system enables us to take in food needed for growth.
☐ The human body is made up of a number of different systems.
☐ The endocrine system consists of various glands, such as the thyroid, sex, and adrenal glands.

(2) ☐ The function of these glands is to secrete chemicals, known as hormones, into the blood.
☐ The endocrine system consists of various glands, such as the thyroid, sex, and adrenal glands.
☐ These hormones control various processes in the body, such as growth, sexual activities, and digestion.

(3) ☐ The process by which they are decomposed is called electrolysis.
☐ Some liquids which act as conductors of electricity decompose when an electric current passes through them.
☐ Such liquids, usually solutions of certain chemicals in water, are known as electrolytes.

(4) ☐ A solid substance can be changed into a liquid substance.
☐ Steam may be converted into water and water into ice.
☐ All substances, except those which decompose when heated, can be changed from one state to another.
☐ A substance in the gaseous state may be changed into a liquid substance.

continued

continued

(5) ☐ Bases are a class of substances, consisting mainly of oxides and hydroxides of metals, which will neutralise acids if used in proper quantities and form salt-like substances in the process.

☐ Compounds can be divided into a number of classes, the most important of which are acids, bases, and salts.

☐ An acid will turn litmus red and it will react with washing soda, giving off carbon dioxide.

☐ A salt is always one of the products when an acid is neutralised with a base or a metal is dissolved by an acid.

(6) ☐ A special class of bases, called alkalis, will dissolve in water and form solutions which turn red litmus blue.

☐ Bases are a class of substances, consisting mainly of oxides and hydroxides of metals, which will neutralise acids if used in proper quantities and form salt-like substances in the process.

☐ Alkalis form solutions which feel soapy and which will dissolve oily and greasy substances, and for this reason they are frequently used for cleaning.

(7) ☐ Problem solving was chosen as the theme for the 1980's by the National Council of Teachers of Mathematics.

☐ There is much academic literature on the subject of problem solving to which artificial intelligence researchers have made significant contributions.

☐ Problem solving is not an activity which is the exclusive property of mathematics; nevertheless, it is a topic which frequently emerges within the mathematical domain.

Practice session 2

Identify the topic sentences in the following paragraphs. (If you cannot find a topic sentence, try to write your own topic sentence for the paragraph by asking yourself the question: 'What is the main idea of this paragraph? What is it about?') Check your answers against those at the end of this chapter before continuing to the next section.

(1) A mysterious plague among the owls at London Zoo has recently claimed 55 victims from 21 different species. Experts have finally identified the killer as an insecticide which reached the owls by a very unusual route. The ones that died were fed mostly on mice. The mice had been reared in cages with sawdust bedding and the sawdust had come from a builder who was using a wood preservative containing the organo-chlorine insecticide dieldrin. The mice absorbed enough from their bedding to kill the owls to which they were fed.

Bolitho, A R & P L Sandler
Study English for Science Longman 1980 54

continued

———— *continued* ————

Topic Sentence / Main Idea: ....

....

....

(2) A food web can be regarded as a number of connected food chains. For example, the herring eats many different animals and therefore forms the top of many different food chains. Again, we find that the animals that are eaten by the herring are also eaten by other top carnivores. Clearly, any particular species is likely to belong to many food chains, as can be seen from Figure 4.2 on page 43, which shows the feeding relationships of the herring. This diagram represents only a very small part of the food web which includes these particular organisms.

Pearson, Ian
English in Biological Science Oxford University Press 1978 40

Topic Sentence / Main Idea: ....

....

....

(3) Some of the properties of magnets were known from very early times. For example, it was known over 2 000 years ago that the mineral magnetite, an oxide of iron, possesses the property of attracting iron. The Chinese, earlier than 2 500 BC, knew that if a piece of magnetite is suspended so that it can turn freely in a horizontal plane, it will set in a definite direction and can therefore be used as a primitive compass. Later it was found that if a bar of iron is rubbed with a piece of magnetite, or lodestone, the magnetic properties of the lodestone are transferred to the iron. The lodestone is called a natural magnet as distinct from other types of magnet which are made by various artificial processes.

Allen, J B & H G Widdowson
English in Physical Science Oxford University Press 1979 65

Topic Sentence / Main Idea: ....

....

....

(4) Acids can be classified into two groups. Acids which always contain the element carbon are called organic acids and they often come from growing things, like fruit. Citric acid, which is found in lemons and oranges and other citrus fruits, and acetic acid, which is found in vinegar, are organic acids. Acids which do not contain the element carbon are known as inorganic acids. They are usually prepared from non-living matter. Inorganic acis consist only of hydrogen and an acid radical. Hydrochloric acid consists of hydrogen and the chloride radical, and sulphuric acid consists of hydrogen and the sulphate radical. They are inorganic acids.

Allen, J B & H G Widdowson
English in Physical Science Oxford University Press 1979 11

———— *continued* ————

continued

Topic Sentence / Main Idea: ....

....

....

(5) When an inland sea evaporates, the solid that is deposited is not a uniform mixture. Instead, because the least soluble salts are deposited first, followed by the more
soluble ones, a typical salt deposit consists of several layers. Each of these layers
consists of a relatively pure salt such as rock salt, NaCl, sylvite, KC1, and carnallite, $KMgC1_3.6 H_2O$. These salt deposits are mined either by conventional methods
or by forcing water into them to form a concentrated salt solution, brine, which
is pumped to the surface and evaporated.

Gillespie, R J et al
Chemistry Allyn & Bacon 1986 560.

Topic Sentence/Main Idea: ....

....

....

(6) In general a bug is some structural flaw (faulty part) manifested in faulty behaviour
(a process). Thus the term 'bug' is used to refer to the incorrect part of a constructed
procedure (for example, incorrect statement in a computer program). It also refers
to an incorrect inference procedure (for example, error in student's subtraction procedure) and, by extension, an error in the student's general model (for example, that
$9 – 4 = 6$). An incorrect general model is commonly called a 'misconception' (such
as misconception about the cause of a disease). Errors can be combined, because
an incorrect computer program may involve a combination of a misconception about
the operators of the computer language and an error in the inference procedure
by which the student has pieced together these operators to accomplish some goal
(that is, how he writes a program). There are, thus, a number of sources of error
possible in the process of cognitive modelling involved in computer programming.

Self, John (ed)
Artificial Intelligence and Human Learning Chapman and Hall 1988 61

Topic Sentence/Main Idea: ....

....

....

Check the answers at the end of this chapter before continuing any further.

☐ Supporting Sentences

Each paragraph thus has a main idea, usually expressed in a topic sentence. The other sentences in the paragraph, then, are *supporting sentences* – sentences which support the main idea by giving examples, adding details, providing reasons, giving facts, and so on.

Examples

☐ Topic Sentence: Some of the properties of magnets were known from very early times.

Supporting Sentence
(giving an example): For example, it was known over 2 000 years ago that the mineral magnetite, an oxide of iron, possesses the property of attracting iron.

Supporting Sentences
(giving facts): The Chinese, earlier than 2 500 BC, knew that if a piece of magnetite is suspended so that it can turn freely in a horizontal plane, it will set in a definite direction and can therefore be used as a primitive compass. Later it was found that if a bar of iron is rubbed with a piece of magnetite, or lodestone, the magnetic properties of the lodestone are transferred to the iron.

☐ Topic Sentence: A mysterious plague among the owls at London Zoo has recently claimed 55 victims from 21 different species.

Supporting Sentences
(providing details): Experts have finally identified the killer as an insecticide which reached the owls by a very unusual route. The ones that died were fed mostly on mice. The mice had been reared in cages with sawdust bedding and the sawdust had come from a builder who was using a wood preservative containing the organo-chlorine insecticide dieldrin. The mice absorbed enough from their bedding to kill the owls to which they were fed.

☐ Topic Sentence: When an inland sea evaporates, the solid that is deposited is not a uniform mixture.

Supporting Sentence
(giving a reason):   Instead, because the least soluble salts are deposited first, followed by the more soluble ones, a typical salt deposit consists of several layers.

Supporting Sentence
(giving details):   Each of these layers consists of a relatively pure salt such as rock salt, $NaCl$, sylvite, KCl, and carnallite, $KMgCl_3.6\ H_2O$.

Sometimes, a paragraph may contain, in addition to the topic sentence, sentences which express *sub-topics* of the main idea. Supporting sentences then are usually arranged around the sub-topics.

☐ Topic Sentence: Acids can be classified into two groups.

Sub-topic 1:   Acids which always contain the element carbon are called organic acids and they often come from growing things, like fruit.

Supporting Sentence
(giving examples):   Citric acid, which is found in lemons and oranges and other citrus fruits, and acetic acid, which is found in vinegar, are organic acids.

Sub-topic 2:   Acids which do not contain the element carbon are known as inorganic acids.

Supporting Sentences
(giving details):   They are usually prepared from non-living matter. Inorganic acids consist only of hydrogen and an acid radical.

Supporting Sentences
(giving examples):   Hydrochloric acid consists of hydrogen and the chloride radical, and sulphuric acid consists of hydrogen and the sulphate radical. They are inorganic acids.

Allen, J B & H G Widdowson
English in Physical Science Oxford University Press 1979 11

☐ Paragraph Patterns

The sentences (topic and supporting) in a paragraph can be arranged or ordered in different patterns. The four most common paragraph patterns are:

☐ from general to particular
☐ from particular to general
☐ from whole to parts
☐ from question to answer, or effect to cause

General to particular

This kind of paragraph begins with the main idea, or topic sentence, and then goes on to explain, illustrate, or add details to the main idea. The paragraph may end with a concluding sentence, which sums up or re-states the main idea.

Example

[Topic Sentence] In early times measurements were made by comparing things with parts of the human body. *[Supporting Sentences]* Early units of measurement included the distance from the elbow to the fingers, the width of the hand and the width of the fingers. Some of these human measurements are still used. For example, the inch is based on the length of half the thumb. A foot was originally the length of a man's foot. A mile was one thousand walking steps. *[Concluding Sentence]* These units were only approximate because their standard – the human body – was not constant.

Bates, Martin & Tony Dudley-Evans
General Science Longman 1982 37

Particular to general

This pattern is obviously the reverse of the general to the particular pattern. The paragraph begins with particulars, details, or examples, (the supporting sentences) and leads up to the most general statement (topic sentence or main idea) at the end of the paragraph.

Example

[Subtopic Sentence] In general, a bug is some structural flaw (faulty part) manifested in faulty behaviour (a process). *[Supporting Sentences]* Thus the term 'bug' is used to refer to the incorrect part of a constructed procedure (eg incorrect statement in a computer program). It also refers to an incorrect inference procedure (eg error in student's subtraction procedure) and, by extension, an error in the student's general model (eg believing that 9 − 4 = 6). An incorrect general model is commonly called a 'misconception' (such as misconception about the cause of a disease). Errors can be combined, because an incorrect computer program may involve a combination of a misconception about the operators of the computer language and an error in the inference procedure by which the student has pieced together these operators to accomplish some goal (*ie* how he writes a program). *[Topic Sentence]* There are, thus, a number of sources of error possible in the process of cognitive modelling involved in computer programming.

Self, John (ed)
Artificial Intelligence and Human Learning Chapman and Hall 1988 61

Whole to parts

If the purpose of a paragraph is to list or identify the parts or divisions of a topic, the whole to parts pattern will probably be used. This pattern starts off by indicating the topic, and the remaining sentences will enumerate the parts or stages of the topic.

Example

[Topic Sentence] Metals have certain properties which distinguish them from other types of solids. *[Supporting Sentences]* They are good conductors of heat. Metals are ductile (capable of being drawn out into wire) and malleable (capable of being hammered into thin sheets). Finally, metal surfaces are good reflectors of light. Most metals have a silvery white colour, indicating that light of all wavelengths is being reflected. Gold and copper absorb some light in the blue region and so appear yellow (gold) or red (copper).

Masterton, W L *et al*
Chemistry Principles Saunders 1981 263

The above paragraph has the topic 'properties', and the supporting sentences list or enumerate these properties.

Question to answer; effect to cause

This pattern starts off the paragraph with a question, which the supporting sentences go on to answer. Alternatively, the paragraph may begin with an effect or a result, and the supporting sentences will explain the causes or reasons for this effect.

Examples

[Effect] A mysterious plague among the owls at London Zoo has recently claimed 55 victims from 21 different species. *[Causes]* Experts have finally identified the killer as an insecticide which reached the owls by a very unusual route. The ones that died were fed mostly on mice. The mice had been reared in cages with sawdust bedding and the sawdust had come from a builder who was using a wood preservative containing the organochlorine insecticide dieldrin. The mice absorbed enough from their bedding to kill the owls to which they were fed.

Bolitho, A R & P L Sandler
Study English for Science Longman 1980 54

[Question] How can this rotation of the plane of polarized light – this optical activity – be detected? *[Answer]* It is both detected and measured by an instrument called the polarimeter. It consists of a light source, two lenses (Polaroid or Nicol), and between the lenses a tube to hold the substance that is being examined for optical activity. These are arranged so that the light passes through one of the lenses (polarizer), then the tube, then the second lens (analyzer).

Morrison, RT & RN Boyd
Organic Chemistry Longman 1980 54

☐ Summary

This chapter has introduced the study of paragraphs, and examined their structure and function. The following concepts are important.

A paragraph is a collection of two or more sentences dealing with a single idea, concept, or stage in the development of an argument.

There are four main types of paragraphs:
- ☐ introductory paragraphs
- ☐ expository paragraphs
- ☐ transitional paragraphs
- ☐ concluding paragraphs

Paragraphs revolve around a single main idea, which may be expressed in a topic sentence – a sentence which summarises the central concern or thesis of the paragraph.

Supporting sentences are the sentences in a paragraph which explain, illustrate, or add details and particulars to the topic sentence or main idea of the paragraph.

Paragraphs are arranged in patterns, the most common of which are the following:
- ☐ general to particular
- ☐ particular to general
- ☐ whole to parts
- ☐ question to answer; effect to cause

In the next chapter, we will look at signpost words – words which help you to discriminate between main ideas and supporting details, and assist you in tracking through a writer's argument.

☐ Answers

Practice session 1

(1) The human body is made up of a number of different systems.

(2) The endocrine system consists of various glands, such as the thyroid, sex, and adrenal glands.

(3) Some liquids which act as conductors of electricity decompose when an electric current passes through them.

———— *continued* ————

─── *continued* ───

(4) All substances, except those which decompose when heated, can be changed from one state to another.

(5) Compounds can be divided into a number of classes, the most important of which are acids, bases, and salts.

(6) Bases are a class of substances, consisting mainly of oxides and hydroxides of metals, which will neutralise acids if used in proper quantities and form salt-like substances in the process.

(7) Problem solving is not an activity which is the exclusive property of mathematics; nevertheless, it is a topic which frequently emerges within the mathematical domain.

Practice session 2 – Topic Sentences

(1) A mysterious plague among the owls at London Zoo has recently claimed 55 victims from 21 different species.

(2) A food web can be regarded as a number of connected food chains.

(3) Some of the properties of magnets were known from very early times.

(4) Acids can be classified into two groups.

(5) When an inland sea evaporates, the solid that is deposited is not a uniform mixture.

(6) There are, thus, a number of sources of error possible in the process of cognitive modelling involved in computer programming.

5
Signpost Words

The last chapter dealt with different kinds of paragraphs and with their construction. We focussed particularly on distinguishing main ideas from supporting detail. Tracking your way through a writer's argument, sorting out the crucial material from the decorative detail, and coming to grips with how various parts of a subject fit together are not tasks that are left solely to you as a guessing game. Writers provide you with 'signpost' words, which guide you through your reading, and in this chapter we are going to study these signpost words so that you are better equipped to understand the texts you read.

Definition

> Signpost words are words used by a writer to act as cues or directions for readers to guide them through the ideas in the text in order to arrive at a full and coherent understanding of the content.

Example

One source of evidence that species evolve is Comparative Embryology. When we compare the embryos of different vertebrates, we find that there are some obvious similarities. *However,* *if* the adult forms are then compared, many of these similarities cannot be found. *For example,* all vertebrate embryos possess very similar embryonic bones just behind the brain-case. In a fish these bones develop into jaw bones, *whereas* in a mammal they develop into ear-bones and function as part of the hearing apparatus. *If* all vertebrates have a common ancestor, we can begin to explain why the embryos are similar.

Pearson, Ian
English in Biological Science Oxford University Press 1978 82

Features of signpost words

(1) Signpost words are usually found at the beginning of a sentence or paragraph, or after a punctuation mark within a sentence.

(2) A signpost word at the *beginning of a sentence* usually indicates how the idea in that sentence is linked to the idea in the preceding sentence, or, alternatively, to the main idea of the paragraph.

Example

A radiator is usually rectangular in cross-section, with wide, flat sides. This shape gives it a large surface area in proportion to its volume. *Consequently*, it gives out more heat.

Here, the signpost word 'consequently' shows that the sentence that follows will deal with a result (or consequence) of the factors mentioned in the preceding sentence.

(3) A signpost word *within a sentence* is used to indicate the relationship between the idea in the second part of the sentence and the idea in the first part of the sentence.

Example

The endocrine system consists of various glands, *such as* the thyroid, sex, and adrenal glands.

In this sentence, the signpost words 'such as' indicate that the second part of the sentence will provide examples of the concept mentioned in the first part of the sentence.

(4) A signpost word at the *beginning of a paragraph* may be used to indicate how the ideas in that paragraph are going to be used to advance the general argument of the text. Signpost words at the beginning of a paragraph can also indicate what kind of paragraph is to follow – an introductory, concluding, transitional or expository paragraph.

Example

In addition to the conceptual problems which Computer Assisted Learning faces, there are a range of pragmatic issues which few researchers have begun to tackle. These involve a consideration of current educational needs, the level of computer provision, teacher expertise, and issues concerning the processes of innovation in education. ...

The signpost words 'in addition' indicate that this paragraph is going to add new information to what has already been given about the topic in preceding paragraphs. It is an expository paragraph, providing new data on the subject being discussed.

(5) Signpost words are used to connect ideas. A variety of connections is possible, and different signpost words are used for each kind of connection.

☐ Kinds of Signpost Words

Additive words

Additive signpost words, obviously, add information to what has already been said. Additive words also imply that the new information is as important and significant as the preceding information.

also	further
and	furthermore
as well as	in addition
at the same time	likewise
besides	moreover
equally important	too

Example

In addition to the conceptual problems which Computer Assisted Learning faces, there are a range of pragmatic issues which few researchers have begun to tackle. These involve a consideration of current educational needs, the level of computer provision, teacher expertise, *and* issues concerning the processes of innovation in education. ...

Practice session 1

Underline the additive signpost words in the following paragraphs.

(1) Orientation of chlorination shows that chlorine atoms, like bromine atoms, preferentially attack benzylic hydrogen; but, as we see, the preference is less marked. Furthermore, competition experiments show that, under conditions where $3°$, $2°$, and $1°$ hydrogens show relative reactivities of $5.0 : 3.8 : 1.0$, the relative rate per benzylic hydrogen of toluene is only 1.3.

Morrison, R T & R N Boyd
Organic Chemistry Allyn & Bacon 1983 642

(2) The Lewis structure of the planar triangular carbonate ion indicates that the central carbon atom forms one double bond and two single bonds to oxygen atoms. We therefore expect one of the bonds in this ion to be shorter than the other two and two of the angles to be larger than the third. However, determination of the structure of this ion by X-ray crystallography has shown that all three bonds have the same length and that the bond angles are all exactly $120°$. Similarly, the Lewis structures of the phosphate and sulfate ions have both single and double bonds. Yet again, researchers have found experimentally that all the bonds are the same length and all the bond angles are equal; these ions have a regular tetrahedral shape.

Gillespie, R J *et al*
Chemistry Allyn & Bacon 1986 300

Answers are at the end of the chapter.

Amplification words

Amplification signpost words introduce information that is to 'amplify' (*ie* expand or enlarge upon) the preceding ideas in the text by giving specific instances. These signpost words are usually followed by examples, illustrations, or concrete instances of the ideas.

as	specifically
for example	such as
for instance	that is
in fact	to illustrate

Example

The endocrine system consists of various glands, *such as* the thyroid, sex, and adrenal glands.

Practice session 2

Underline the amplification signpost words in the following paragraphs.

(1) Substances consist of small parts, or particles, which are known as molecules. Molecules are composed of atoms. Some substances, like salt and water, have molecules which can be analysed further into other substances. If a molecule of water is analysed, for example, it will be found to consist of two atoms of hydrogen and one atom of oxygen. Substances whose molecules are composed of atoms of other substances are known as compounds. Other substances have molecules which cannot be broken down into atoms of other substances, and these are called elements. Hydrogen and oxygen, for example, are elements.

Allen, J P B & H G Widdowson
English in Physical Science Oxford University Press 1974 22

(2) Food contains only minute quantities of the substances called vitamins, but they are vital for good health. For example, if you eat a diet of meat, bread, sugar and fat, you may become ill with a disease called scurvy. This is caused by a deficiency in vitamin C, which is found in fruit and vegetables.

Bates, Martin & Tony Dudley-Evans
General Science Longman 1990 72

(3) A substance may be an element, a compound, or a mixture. An element, such as nitrogen or iron, cannot be broken down into simpler substances. When two or more elements combine, they form a compound.

Bates, Martin & Tony Dudley-Evans
General Science Longman 1990 10

Answers are at the end of this chapter.

Repetitive words

Repetitive signpost words introduce a re-statement of an idea or concept. The writer may repeat something in different words in order to make the concept clearer, and to emphasise its importance.

| again | that is (*ie*) |
| in other words | to repeat |

Practice session 3

Underline the repetitive signpost words in the following paragraph.

(1) Almost all stable molecules have an even number of electrons. In other words, all their electrons are in filled orbitals, and they are therefore much less reactive than free atoms. Molecules containing an odd number of electrons are known, but like most free atoms, they are almost all extremely reactive, because they have an incompletely filled orbital – that is, an unpaired electron.

Gillespie, R J et al
Chemistry Allyn & Bacon 1986 230

Answers are at the end of this chapter.

Contrast and change words

Contrast and change signpost words are used to introduce 'the other side of the story'. The writer will probably have presented you with one set of facts, circumstances or ideas, and now changes track to present you with alternative or contrasting data.

but	notwithstanding
conversely	on the other hand
despite	still
even though	though
however	whereas
in contrast	yet

Practice session 4

Underline the contrast and change words in the following passages.

(1) The climate and vegetation were still similar to those of the Eocene, and Mesohippus also fed on broad-leaved plants. However, by the middle of the Miocene, the climate was drier and soft grasses replaced the broad-leaved plants.

Pearson, Ian
English in Biological Science Oxford University Press 1978 92

(2) Among these 'other-feeders' or heterotrophs, we distinguish between 'solid-feeders' or phagotrophs, and 'liquid-feeders', or osmotrophs. Whereas phagotrophic organisms take in solid and often living food, osmotrophic ones absorb or suck up liquid food.

Pearson, Ian
English in Biological Science Oxford University Press 1978 13

continued

───── *continued* ─────

(3) In a general way, we can relate the energy change in a reaction to the difference in energy between products and reactants. If the products have a higher energy than the reactants, we must supply energy to make the reaction proceed. Conversely, if the products are in a lower energy state than the reactants, we can get energy out of the reaction.

Masterton, W L et al
Chemistry Principles Saunders College 1981 106

(4) The Lewis structure of the planar triangular carbonate ion indicates that the central carbon atom forms one double bond and two single bonds to oxygen atoms. We therefore expect one of the bonds in this ion to be shorter than the other two and two of the angles to be larger than the third. However, determination of the structure of this ion by X-ray crystallography has shown that all three bonds have the same length and that the bond angles are all exactly 120°. Similarly, the Lewis structures of the phosphate and sulfate ions have both single and double bonds. Yet again, researchers have found experimentally that all the bonds are the same length and all the bond angles are equal; these ions have a regular tetrahedral shape.

Gillespie, R J et al
Chemistry Allyn & Bacon 1986 300

(5) Orientation of chlorination shows that chlorine atoms, like bromine atoms, preferentially attack benzylic hydrogen; but, as we see, the preference is less marked. Furthermore, competition experiments show that, under conditions where 3°, 2°, and 1° hydrogens show relative reactivities of 5.0: 3.8: 1.0, the relative rate per benzylic hydrogen of toluene is only 1.3.

Morrison, R T & R N Boyd
Organic Chemistry Allyn & Bacon 1983 642

Cause and effect words   Cause and effect signpost words are used to introduce and link ideas of causality and consequence. The words may be used to introduce the reasons or causes for something, or to list the results or effects.

accordingly	since
as a result	so
because	then
consequently	therefore
for this reason	thus

Example

A radiator is usually rectangular in cross-section, with wide, flat sides. This shape gives it a large surface area in proportion to its volume. *Consequently*, it gives out more heat.

Practice session 5

Underline the cause and effect signpost words in the following paragraphs.

(1) Let us consider an example of a process which illustrates the relationship between acids, bases and salts. If a solution of sodium hydroxide is poured into a beaker containing hydrochloric acid and red litmus, the litmus will turn blue. This shows that the acid has been neutralized and that sodium hydroxide is a base. As a result of the neutralization of the acid by the base, sodium chloride is formed and may be obtained as crystals by evaporation of the solution. Thus the result of adding the base sodium hydroxide to hydrochloric acid is to form a salt, sodium chloride.

Allen, J P B & H G Widdowson
English in Physical Science Oxford University Press 1979 23

(2) The Lewis structure of the planar triangular carbonate ion indicates that the central carbon atom forms one double bond and two single bonds to oxygen atoms. We therefore expect one of the bonds in this ion to be shorter than the other two and two of the angles to be larger than the third. However, determination of the structure of this ion by X-ray crystallography has shown that all three bonds have the same length and that the bond angles are all exactly 120°. Similarly, the Lewis structures of the phosphate and sulfate ions have both single and double bonds. Yet again, researchers have found experimentally that all the bonds are the same length and all the bond angles are equal; these ions have a regular tetrahedral shape.

Gillespie, R J *et al*
Chemistry Allyn & Bacon 1986 300

(3) Heat causes substances to expand. This is because heat causes the atoms and molecules in the substance to move more quickly. As a consequence, they take up more space. This is true for gases, liquids, and solids, but gases expand much more than liquids, and liquids much more than solids. When a substance is cooled, the molecules slow down and as a result the substance contracts.

Bates, Martin & Tony Dudley-Evans
General Science Longman 1990 84

Check your answers against those at the end of the chapter.

Qualifying words

Qualifying signpost words indicate the conditions under which the ideas, concepts, or facts are valid or are to be considered. Qualifying words are important because they introduce information which is usually a crucial prerequisite to the validity of the data or concept under discussion.

although	providing
if	unless

Practice session 6

Underline the qualifying signpost words in the following paragraphs.

(1) If we put some dilute sulphuric acid in a beaker and dip a plate of pure zinc in it, there is no visible action. If we dip a plate of copper into the acid, but without allowing it to touch the zinc plate, still no action will be observed. However, if we allow the zinc and copper plates to touch one another, at once bubbles of hydrogen will rise from the copper plate.

> Allen, J P B & H G Widdowson
> *English in Physical Science* Oxford University Press 1979 23

(2) The activating energy is provided by the kinetic energy of the reacting molecules. Unless the kinetic energy of their relative motion is at least equal to E_a, no reaction will occur, and the molecules simply bounce apart unchanged. The activation energy represents a barrier that must be overcome if the reaction is to occur.

> Gillespie, R J *et al*
> *Chemistry* Allyn & Bacon 1986 662

Consult the answers at the end of this chapter.

Emphasis-ing words

Emphasising signpost words are used to highlight and underline important points that the writer wishes to drive home. They are a cue for you to wake up and take notice!

above all	more / most important(ly)
	more / most significant(ly)

Practice session 7

Underline the emphasizing signpost words in the following paragraph.

(1) A chemical equation is a shorthand description of a reaction that gives the formulas for all the reactants and all the products. For example,

Reactants		Products
$2 H_2 + O_2$	$\longrightarrow$	$2 H_2O$
$C + O_2$	$\longrightarrow$	CO_2
$CH_4 + 2O_2$	$\longrightarrow$	$CO_2 + 2 H_2O$

Most importantly, an equation must be consistent with the law of conservation of mass, which tells us that atoms are neither created nor destroyed in chemical reactions.

> Gillespie, R J *et al*
> *Chemistry* Allyn & Bacon 1986 63

Order words

Order signpost words are used to mark out a chronology of events, or to list data in a specific sequence.

afterwards	now
at the same time	presently / today
before	subsequently
first(ly), second(ly) ... etc	then
formerly	ultimately
last(ly)	until
later	while
meanwhile	historically
next	historical periods

Note the crucial role of order signpost words in the description of the life-cycle of a plant, for example.

> *First* the seed is sown. *Next*, it is watered. *Then*, the seed begins to swell. *At this stage*, germination begins. *Subsequently*, the roots develop. *Meanwhile*, the leaves also develop. *Later*, flowers appear. *Then*, pollination takes place. *During this process*, the stigma receives pollen. *Afterwards*, the fruit forms. *Eventually*, the plant dies. *Finally*, the plant decomposes.

Bates, Martin & Tony Dudley-Evans
General Science Longman 1982 57

Practice session 8

Underline the order words in the following paragraphs.

(1) Let us see how we arrive at structure I from the experimental facts. First of all, the initial oxidation labels (with a -COOH group) the D-glucose unit that contains the 'free' aldehyde group. Next, methylation labels (as -OCH$_3$) every free -OH group. Finally, upon hydrolysis, the absence of a methoxyl group shows which -OH groups were *not* free.

Morrison, R T & R N Boyd
Organic Chemistry Allyn & Bacon 1983 1099

(2) Although it is not abundant, lead is easily extracted from its most important ore, galena, PbS. As a result, it has been known for a long time. In Ancient Rome lead pipes were used for the water supply of villas; buildings such as the Pantheon had roofs sheathed in lead; and even wine casks were lined with lead. Some historians have claimed that the use of lead by Romans was so extensive that many Romans, particularly the rich governing class, suffered from lead poisoning, which lead to both madness and sterility, and that lead poisoning was one cause of the decline of the Roman Empire. In the Middle Ages lead was used extensively in the building of the great cathedrals, for roofs, gutters, and stained glass windows.

— *continued* —

──── *continued* ────

Today lead is used for making storage batteries, for covering electric cables, and as an important component of several alloys such as type metal and solder. Large quantities are also used for the production of the gasoline additive tetra-ethyl lead, $Pb(C_2H_5)_4$. This use is now declining, since the health hazards associated with 'leaded' gasoline have been recognized.

Gillespie, R J *et al*
Chemistry Allyn & Bacon 1986 314

(3) Eohippus appeared in the early Eocene, when the ground was wet and soft. It was 12 to 20 inches high and had four digits on its forelimb. There was also one splint bone, representing the first digit. It fed on the broad-leaved plants that grew at the time.

Mesohippus evolved about 35 millions of years ago. It had only three digits on its forelimb, although the fifth was represented by a splint bone. The climate and vegetation were still similar to those of the Eocene, and Mesohippus also fed on broad-leaved plants. However, by the middle of the Miocene the climate was drier and soft grasses replaced the broad-leaved plants.

Pearson, Ian
English in Biological Science Oxford University Press 1978 91

Check your answers against those at the end of this chapter.

Summarising words

Writers use summarising words to introduce a recapitulation of the data in a condensed or summarised form. They usually do this at a stage in their argument when it is important that you have grasped the essentials before proceeding to deal with new information. A summarising signpost word is thus your cue to read carefully to check your understanding of what you have read so far, and perhaps to mark the sentence or paragraph with a highlighting pen as a useful précis.

briefly	in conclusion
in brief	to summarise / to sum up
in short	in summary

Practice session 9

Underline the summarising signpost words in the following paragraphs.

(1) It is now clear why we talk of natural selection. In brief, we mean by this phrase that nature (in other words, the environment) selects, or chooses, the individuals which will reproduce and so pass on their characteristics. These selected individuals are the ones that are best adapted to their environment. Because the environment may alter, and because individuals may move to places where the environment is

──── *continued* ────

continued

different, survival may suddenly come to depend upon quite different characteristics. In such circumstances, descent with modification is then the key to the survival of the species.

Pearson, Ian
English in Biological Science Oxford University Press 1978 93

(2) In summary, the melting points of molecular solids are usually low, often well below room temperature and always below 400°C. In contrast, network solids usually have high melting points, often as high as several thousand degrees; with very few exceptions network solids melt above room temperature.

Gillespie, R J *et al*
Chemistry Allyn & Bacon 1986 361

Check your answers against those at the end of the chapter.

Why are signpost words important?

Signpost words direct your reading and your thinking. They also point to ideas, modify or connect ideas, and even reverse ideas in mid-sentence. Just as you would never find your way in an unfamiliar part of town without signposts giving street names, so without signpost words to guide you through unfamiliar data or new ideas, you will become lost in a muddle of facts, without any clear understanding or integration of the material.

Example

The way in which signpost words help you to identify the topic sentence of a paragraph, and understand how the supporting sentences relate to the topic sentence can been seen in the following example.

A food web can be regarded as a number of connected food chains. *For example*, the herring eats many different animals and therefore forms the top of many different food chains. *Again*, we find that the animals that are eaten by the herring are also eaten by other top carnivores. *Clearly,* any particular species is likely to belong to many food chains, as can be seen from Figure 4.2 on page 43, which shows the feeding relationships of the herring. This diagram represents only a very small part of the food web which includes these particular organisms.

Pearson, Ian
English in Biological Science Oxford University Press 1978 40

The first sentence must be the topic sentence, as the second sentence is an example introduced by an amplification signpost ('for example'). The third sentence is an additional detail introduced by a repetitive signpost word ('again'), and the fourth sentence draws a conclusion, introduced by the signpost word 'clearly'.

How to use signpost words

Initially, you should consciously try to notice and mark the signpost words in a body of text, reminding yourself of the function of each. As your reading becomes more proficient, you will find that the signpost words and their significance have a subconscious effect on your understanding and even the most dense and complex texts become more accessible.

☐ Answers

Practice session 1 – Additive Signpost Words

(1) ... the preference is less marked. *Furthermore,* competition experiments show ... .

(2) ... the central carbon atom forms one double bond *and* two single bonds to oxygen atoms. We therefore expect one of the bonds in this ion to be shorter than the other two *and* two of the angles to be larger than the third. ... all three bonds have the same length *and* that the bond angles are all exactly 120°. *Similarly,* the Lewis structures of the phosphate and sulfate ions have both single *and* double bonds. *Yet again,* researchers have found experimentally that all the bonds are the same length *and* all the bond angles are equal; these ions have a regular tetrahedral shape.

Practice session 2 – Amplification Words

(1) If a molecule of water is analysed, *for example,* it will be found to consist of two atoms of hydrogen and one atom of oxygen. Hydrogen and oxygen, *for example,* are elements.

(2) *For example,* if you eat a diet of meat, bread, sugar and fat, you may become ill with a disease called scurvy.

(3) An element, *such as* nitrogen or iron, cannot be broken down into simpler substances.

Practice session 3 – Repetitive Words

(1) *In other words,* all their electrons are in filled orbitals, and they are therefore much less reactive than free atoms. ... they have an incompletely filled orbital – *that is,* an unpaired electron.

Practice session 4 – Contrast and Change Words

(1) *However,* by the middle of the Miocene, the climate was drier and soft grasses replaced the broad-leaved plants.

(2) *Whereas* phagotrophic organisms take in solid and often living food, osmotrophic ones absorb or suck up liquid food.

(3) *Conversely,* if the products are in a lower energy state than the reactants, we can get energy out of the reaction.

(4) *However,* determination of the structure of this ion by X-ray crystallography has shown that all three bonds have the same length and that the bond angles are all exactly $120°$. ... *Yet* again, researchers have found experimentally that all the bonds are the same length and all the bond angles are equal; these ions have a regular tetrahedral shape.

(5) Orientation of chlorination shows that chlorine atoms, like bromine atoms, preferentially attack benzylic hydrogen; *but,* as we see, the preference is less marked.

Practice session 5 – Cause and Effect Words

(1) *As a result* of the neutralisation of the acid by the base, sodium chloride is formed and may be obtained as crystals by evaporation of the solution. *Thus* the result of adding the base sodium hydroxide to hydrochloric acid is to form a salt, sodium chloride.

(2) We *therefore* expect one of the bonds in this ion to be shorter than the other two and two of the angles to be larger than the third.

(3) Heat causes substances to expand. This is *because* heat causes the atoms and molecules in the substance to move more quickly. *As a consequence,* they take up more space. When a substance is cooled, the molecules slow down and *as a result* the substance contracts.

Practice session 6 – Qualifying Words

(1) *If* we put some dilute sulphuric acid in a beaker and dip a plate of pure zinc in it, there is no visible action. *If* we dip a plate of copper into the acid, but without allowing it to touch the zinc plate, still no action will be observed. However, *if* we allow the zinc and copper plates to touch one another, at once bubbles of hydrogen will rise from the copper plate.

continued

———— *continued* ————

(2) *Unless* the kinetic energy of their relative motion is at least equal to E_a, no reaction will occur, and the molecules simply bounce apart unchanged.

Practice session 7 – Emphasising Words

(1) *Most importantly,* an equation must be consistent with the law of conservation of mass, which tells us that atoms are neither created nor destroyed in chemical reactions.

Practice session 8 – Order Words

(1) ... *First* of all, the initial oxidation ... . *Next,* methylation ... . *Finally,* upon hydrolysis, ...

(2) *In Ancient Rome* lead pipes were used for the water supply of villas ... . *In the Middle Ages* lead was used extensively in the building of the great cathedrals ... . *Today* lead is used for making storage batteries... . This use is *now* declining ... .

(3) Eohippus appeared *in the early Eocene* ... . Mesohippus evolved *about 35 millions of years ago.* ... However, *by the middle of the Miocene* the climate was drier.

Practice session 9 – Summarising Words

(1) It is now clear why we talk of natural selection. *In brief,* we mean by this phrase that nature (in other words, the environment) selects, or chooses, the individuals which will reproduce and so pass on their characteristics.

(2) *In summary,* the melting points of molecular solids are usually low, often well below room temperature and always below 400°C.

6

Types of Paragraphs (1)

☐ Descriptions and Definitions

In the past two chapters, we have been looking at the mechanics of paragraphs – what they consist of, how they are constructed, how they function. This chapter will look at what paragraphs *do*; that is, the purpose and content of the different kinds of paragraphs common to scientific writing. The major function of paragraphs is, obviously, to convey information to the reader. In scientific texts, writers most often convey this information by describing, defining, classifying, or instructing. Thus we tend to find four main types of paragraph in scientific texts:

- ☐ paragraphs of description
- ☐ paragraphs of definition
- ☐ paragraphs of classification
- ☐ paragraphs of instruction.

In this chapter we will examine descriptions and definitions as they appear in scientific texts. Chapter 7 will discuss paragraphs of classification. Paragraphs of instructions are fairly straightforward and are thus not dealt with.

☐ Paragraphs of Description

Definition

A paragraph of description describes physical appearance, function, properties or procedure.

Physical description

(1) Physical description provides information on the physical characteristics of an object. These include-:

dimension	shape	weight
colour	texture	material
volume	structure	density
reactivity	capacity	composition

Examples

Trachelomonas has two flagella. It is egg-shaped and roughly 45 microns in length. It possesses one nucleus, about 10 plastids, a reservoir, and a contractile vacuole.

Pearson, Ian
English in Biological Science Oxford University Press 1978 9

Magnets today are usually made of special alloys of steel. A steel magnet differs from ordinary steel and from all other substances in three important respects: it attracts iron filings, it sets in a definite direction when freely suspended, and it converts iron and steel bars in its neighbourhood into magnets.

Allen, J P B & H G Widdowson
English in Physical Science Oxford University Press 1974 65

(2) Physical description can describe the spatial relationship of the parts of the object to one another and to the whole; and the spatial relationship of the whole object to other objects.

Spatial relationship descriptions make generous use of terms to indicate position, such as 'next to', 'behind', 'perpendicular to', 'in a tetrahedral arrangement', etc. (Note their use in the examples below.)

Examples

Barometers are instruments which are used for measuring atmospheric pressure. The aneroid barometer consists of a thin metal box in the shape of a concertina containing a partial vacuum. The box is prevented from collapsing by means of a steel spring fixed *at the side* of it. The spring is bent *over the top* of the box, exerting pressure *upon* it by means of a metal shaft, one end of which is attached to the spring. The other end is attached to a flat piece of metal which presses *down on the top* of the box. The spring extends *sideways* from the top of the metal shaft and then bends *downwards* to connect up with a short horizontal lever which is connected to a long vertical lever by means of a pivot. The longer lever extends *upwards* and is connected to a thread which passes *sideways* and winds on to the axle of a pointer fixed *above* a dial. A small coiled spring is attached to this axle.

Allen, J P B & H G Widdowson
English in Physical Science Oxford University Press 1974 86

It can be seen from a cross-section of a young root that the cells are arranged into different tissues. The *outermost* layer of cells forms the piliferous layer, and if the root is young enough, it can be seen that most of the cells bear elongated root hairs. *Inside* the piliferous layer there is the exodermis, which becomes the outermost layer when the piliferous layer is lost. *Between* the exodermis and the endodermis there is a broad cortex that is many cells wide. The *last* complete ring of cells is the pericycle. It lies *inside* the endodermis and it *surrounds* the phloem, cambium, and xylem.

Pearson, Ian
English in Biological Science Oxford University Press 1978 9

(3) Physical descriptions may be in the form of diagrams, and descriptive paragraphs are often accompanied by diagrams. These can be considerably more useful than text in helping readers to visualise the object being described, and thus should not be ignored, but 'read' as closely as the text. The two descriptions below illustrate how efficiently a diagram can convey clear information.

Examples

The apparatus for preparing hydrogen consists of a flask, a gas-jar, a beehive shelf, a trough, a delivery tube and a thistle funnel. The flask is spherical and has a flat bottom. It contains zinc and hydrochloric acid. The thistle tube and the delivery tube are fitted into the neck of the flask. They are held in place by a two-holed cork. The thistle tube leads down to the hydrochloric acid. The delivery tube leads from the flask to the hole in the beehive shelf. The beehive shelf is placed in the middle of the trough. The trough contains water. The gas-jar is supported by the beehive shelf. Hydrogen is collected at the top of the gas-jar.

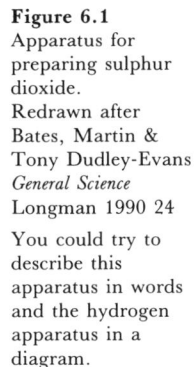

Figure 6.1
Apparatus for preparing sulphur dioxide.
Redrawn after Bates, Martin & Tony Dudley-Evans *General Science* Longman 1990 24

You could try to describe this apparatus in words and the hydrogen apparatus in a diagram.

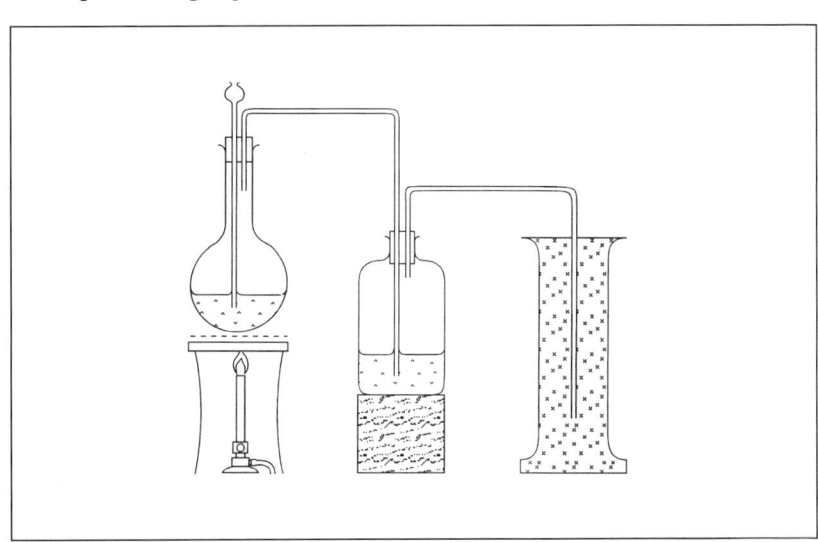

Of the above two examples, the diagram is obviously a much clearer description to 'read' than is the paragraph. Diagrams, pictures, charts, tables, graphs, etc are seldom included in scientific texts for decorative purposes. They are important conveyors of information, and should be given close attention. The example below further illustrates the density and clarity of the information that can be presented in a diagram combined with text. This is taken up again in chapter 13.

Figure 6.2
Transverse section of a leaf
Redrawn after Pearson, Ian *English in Biological Science* Oxford University Press 1978 33

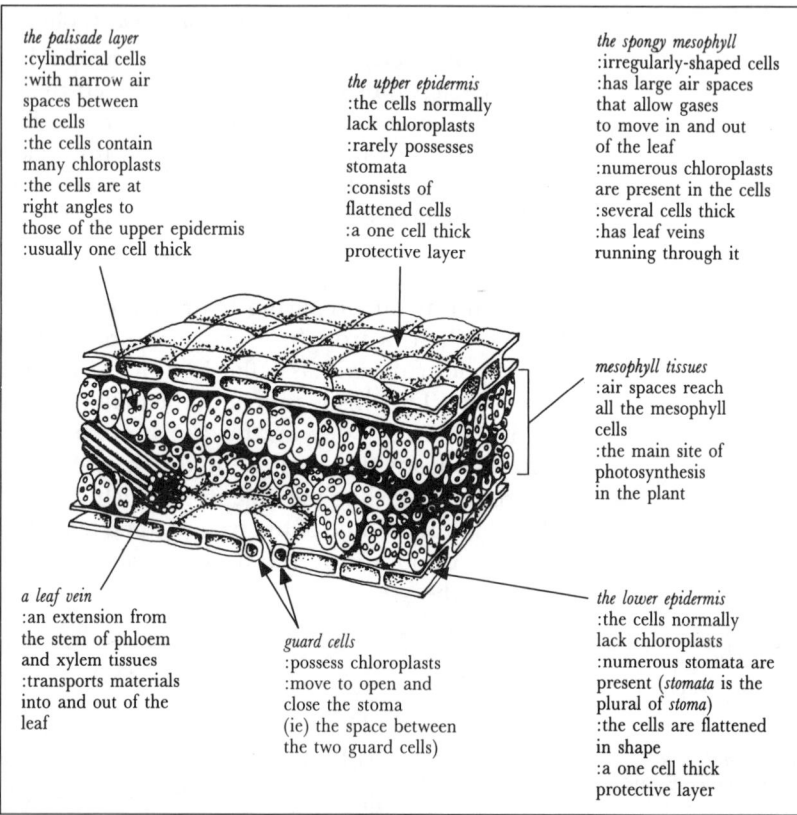

the palisade layer
:cylindrical cells
:with narrow air spaces between the cells
:the cells contain many chloroplasts
:the cells are at right angles to those of the upper epidermis
:usually one cell thick

the upper epidermis
:the cells normally lack chloroplasts
:rarely possesses stomata
:consists of flattened cells
:a one cell thick protective layer

the spongy mesophyll
:irregularly-shaped cells
:has large air spaces that allow gases to move in and out of the leaf
:numerous chloroplasts are present in the cells
:several cells thick
:has leaf veins running through it

mesophyll tissues
:air spaces reach all the mesophyll cells
:the main site of photosynthesis in the plant

a leaf vein
:an extension from the stem of phloem and xylem tissues
:transports materials into and out of the leaf

guard cells
:possess chloroplasts
:move to open and close the stoma
(ie) the space between the two guard cells)

the lower epidermis
:the cells normally lack chloroplasts
:numerous stomata are present (*stomata* is the plural of *stoma*)
:the cells are flattened in shape
:a one cell thick protective layer

A good way to summarise a paragraph of physical description is to attempt to draw an accompanying diagram, using the information given in the text.

Function description

(1) Function description usually provides the reader with information about a device of some kind.

(2) Information may be provided about the use or purpose of the device.

(3) Information is given explaining *how* the main parts of the device function. A description of the parts that a device consists of is, strictly speaking, a physical description. Function description, as the term implies, describes how the device works.

Compare the physical description of a bell given on page 28, to the function description describing its operation, given in the example below.

Example

To operate the bell, the key, or screw, is connected to the positive terminal of the battery, and the copper wire coming from the electromagnet is connected to the negative terminal. *When* the current is switched on, it flows through the key into the spring, passing from there round the coils of the electromagnet and then back to the battery. *As* the current passes through the coils of copper wire, the soft iron cylinders around which it is wound become magnetized. *Consequently,* they attract the armature, causing the head of the striker rod to hit the gong. As the striker hits the gong, the spring to which it is fixed loses contact with the screw, breaking the circuit. The current ceasing to flow, the electromagnet loses its magnetism and the armature, being no longer attracted, is pulled back by the spring. *When* this happens, the spring makes contact with the screw once more, allowing the electric current to pass, again magnetizing the cylinders. These *then* attract the armature, once more pulling the spring away from the screw and breaking the circuit. The whole process is repeated over and over again, causing the head of the striker to vibrate rapidly against the gong, *thus* producing the familiar sound of an electric bell.

Allen, J P B, & H G Widdowson
English in Physical Science Oxford University Press 1974 84

(4) Function descriptions tend to use cause and effect signpost words (to explain how one event or action causes another), as well as order signpost words (to mark the sequence of actions or events). These signpost words have been italicised in the above paragraph. Cause and effect chains in describing the functioning of a device can also be constructed without necessarily using signpost words. Look at the following sentence taken from the paragraph, for example:

The current ceasing to flow, the electromagnet loses its magnetism and the armature, being no longer attracted, is pulled back by the spring.

This can be re-written to make the cause and effect sequences clearer:

- ☐ The current ceases to flow (because the circuit has broken – see preceding sentence).
- ☐ [therefore] The electromagnet loses its magnetism.
- ☐ [as a result] The armature is no longer attracted.
- ☐ [thus] The armature is pulled back by the spring.

> Paragraphs containing function descriptions can be summarised by listing the events consecutively (as in the above example) or in a numbered sequence.

Process description

(1) Process description paragraphs describe a procedure, which may be a natural procedure (such as photosynthesis) or a contrived procedure (such as an experiment designed to test a hypothesis).

(2) The description focuses on the steps or stages in the process, setting these out in a logical (and usually chronological) sequence. Order signpost words are therefore commonly used.

Examples

Water passes by osmosis into the root hairs and *then* into the cells of the piliferous layer. *Next,* it passes through the exodermis and *then* enters the cells of the cortex. From the cortex, the water passes into the endodermis and *then* into the pericycle. *Finally* it reaches the xylem cells, *whence* it is transported up through the plant to the leaves.

Pearson, Ian
English in Biological Science Oxford University Press 1978 32

Two copper plates are connected to a battery, *after* having been carefully weighed. They are *then* placed in a glass vessel containing copper sulphate solution. The current is *then* switched on. *After* about half an hour, the current is switched off and the copper plates are taken out of the solution. *After* they have been dried, they are weighed again. One plate *now* weighs more than before and the other one weighs less than before, and the weight lost by the one is equal to the weight gained by the other.

Allen, J P B & H G Widdowson
English in Physical Science Oxford University Press 1974 71

> Paragraphs describing a process can best be summarised by drawing a flow chart, or by listing a numbered sequence.

Examples

☐ *How a record-player works*

(1) Stylus vibrates in grooves in record.
(2) Vibrations received by pick-up.
(3) Pick-up converts vibrations to electric current.
(4) Current passed on to amplifier.
(5) Amplifier increases power of current.
(6) Current activates loudspeaker.

Bolitho, A R & P L Sandler
Study English for Science Longman 1980 71

☐ *Circulatory System*

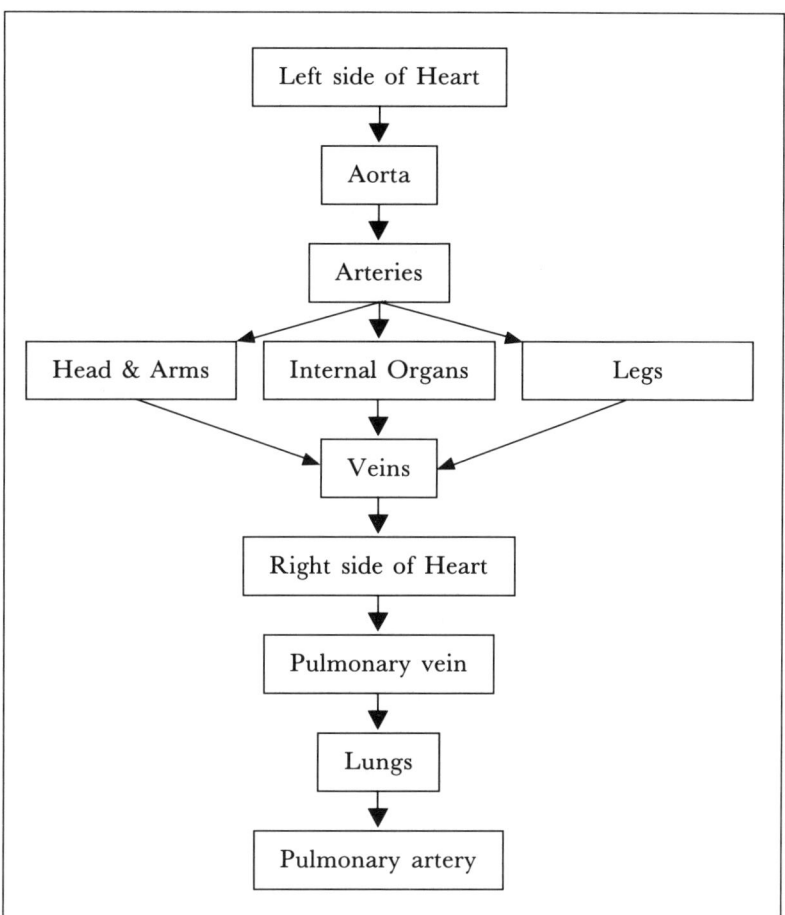

☐ **Paragraphs of Definition**

Definition

A definition is an explanation of the meaning of a word, phrase, concept, or phenomenon.

Writers of scientific texts make frequent use of definitions – which may range from a phrase to paragraph-length explanations. Definitions help the reader to come to grips with new concepts, understand new technology, or grasp a new way of looking at an old idea. In understanding a new concept, it is useful to have the following three pieces of information:

☐ the term [T] being defined
☐ the set, class or category [C] to which the object or phenomenon belongs (or of which the object is a subset)
☐ the characteristics or differences [D] which distinguish the object or phenomenon from others in the same category, or which determine the subset.

Writers can convey this information in a number of ways, some of which are discussed below. As a reader, your job is to be on the lookout for the three basic 'chunks' of information, regardless of the format in which the writer presents them.

Formal definitions

> A formal definition is one which presents the three pieces of information in a rigid format: T → C + D.

Examples

☐ A key is an implement designed to open a lock.

   T : key
   C : implements
   D : designed to open a lock

☐ Hatred is an emotion of violent dislike or animosity.

   T : hatred
   C : emotions
   D : violent
       dislike or animosity

Practice session 1

Analyse the following formal definitions into term [T], class [C], and differences [D]. List the differences separately.

(1) A manticore is a fabulous monster having the head of a man, body of a lion, and the tail of a dragon or scorpion.

   T : ....

   C : ....

   D : ....

   ....

   ....

continued

— *continued* —

(2) Homiletics is the field of study dealing with the art of preaching.

   T : ....

   C : ....

   D : ....

(3) Scotopia is the ability to see in dim light.

   T : ....

   C : ....

   D : ....

   ....

(4) Misogamy is the pathological dislike of marriage.

   T : ....

   C : ....

   D : ....

   ....

(5) A zloty is the basic monetary unit of Poland.

   T : ....

   C : ....

   D : ....

(6) A xyster is a surgical instrument for scraping bones.

   T : ....

   C : ....

   D : ....

(7) Sphragistics is the study of seals and signets.

   T : ....

   C : ....

   D : ....

— *continued* —

—— *continued* ——

(8) A pizza is an Italian baked dish consisting of a shallow pielike crust covered usually with a spiced mixture of tomatoes and cheese.

T : ....

C : ....

D : ....

....

Check your answers against those at the end of this chapter.

Semi-formal definitions   Semi-formal definitions omit one of the three basic pieces of information. The item omitted is often the class, usually because this is considered to be obvious.

Examples

☐ A sphygmomanometer measures blood pressure.

T : sphygmomanometer
C : [instruments]
D : measures blood pressure

☐ A Spenserian sonnet comprises three interlocking quatrains and a couplet with the rhyme pattern abab bcbc cdcd ee.

T : Spenserian sonnet
C : [sonnets]
D : three interlocking quatrains
couplet
rhyme pattern abab bcbc cdcd ee

Practice session 2

Analyse the semi-formal definitions below into T, C, and D. You will need to devise the class or category (as in the above example).

(1) A brake is used for slowing or stopping motion, usually by contact friction.

T : ....

C : ....

D : ....

....

—— *continued* ——

continued

(2) A cold is characterised by inflammation of the mucous membranes of the respiratory passages and accompanying fever, chills, coughing, and sneezing.

T : ....

C : ....

D : ....

....

(3) A pyrotechnist designs, manufactures, and sets off fireworks.

T : ....

C : ....

D : ....

(4) A pencil is narrow and generally cylindrical, consisting of a thin rod of graphite or similar substance, encased in wood, metal or plastic, and is used for writing.

T : ....

C : ....

D : ....

....

(5) An office is where services, clerical work, professional duties, or the like are carried out.

T : ....

C : ....

D : ....

....

√ (6) A diphthong begins with one vowel sound and moves to another vowel or semivowel within the same syllable.

T : ....

C : ....

D : ....

....

continued

—— *continued* ——

(7) An interrobang is a combination of a question mark and an exclamation mark.

T : ....

C : ....

D : ....

(8) Oncology deals with tumours.

T : ....

C : ....

D : ....

....

(9) Nautical miles are based on the length of one minute of arc of a great circle, equivalent to 1 852 metres.

T : ....

C : ....

D : ....

....

(10) Succotash consists of kernels of corn and lima beans cooked together.

T : ....

C : ....

D : ....

....

Check your answers against those at the end of the chapter.

Non-formal definitions

(1) Non-formal definitions do not follow any specific format. They generally provide the term being defined, and words or phrases which have approximately the same meaning of the term, or which indicate an outstanding characteristic.

(2) Non-formal definitions aim at providing the reader with a general sense of a term, rather than with a detailed, specific, and explicit definition. They therefore do not provide as much information as the more formal types of definition.

(3) A common type of non-formal definition is definition by syno-
nym – a word with almost identical meaning to the term being
defined.

Examples

A furuncle is a boil.
An arachnid is a spider.
An eructation is a burp.
A bolero is a jacket.
A blazer is a jacket.
An anorak is a jacket.

The last three examples illustrate the vagueness of this type of
definition – an uninformed reader would not know what distinguishes a bolero from a blazer from an anorak. While ignorance
about items of clothing may not be earthshatteringly important,
in the sciences it is often crucial that your understanding of a concept is accurate and detailed. Should you encounter a definition
as imprecise as those above, go to some trouble to gather the information necessary to re-write it as a formal definition.

(4) Non-formal definitions may also operate by telling you what an
object or phenomenon is not. One way of doing this is by using
antonyms (words which have the opposite meaning to the term
being defined).

Examples

☐ Democracy is the opposite of autocracy.

☐ Consubstantiation is the opposing doctrine to transsubstantiation.

The second example above illustrates the unsatisfactory nature of
definition by antonym. Should you be faced with such a definition, make sure that you understand the antonym!

Definition by negative statement can, however, be a very useful
way of setting the parameters of a term's meaning and applicabil-
ity, particularly if the negative statements are accompanied by positive statement. Note the example below:

Example

A wave equation cannot tell us exactly where an electron is at any particular moment, or how fast it is moving; it does not permit us to plot a
precise orbit about the nucleus. Instead, it tells us the probability of finding
the electron at any particular place.

Morrison, R T & R N Boyd
Organic Chemistry Allyn & Bacon 1983 6

> When dealing with non-formal definitions, make sure that your understanding of the term is complete. Consult other sources if you are unsure.

Expanded definitions

(1) Definitions do not need to be limited to a simple statement of term, class, and one or two differences. They can be expanded by adding additional information to guide the reader towards as complete an understanding of the concept as possible.

(2) Definitions can be expanded by listing additional attributes, parts, or essential characteristics.

Examples

A seal is an aquatic, carnivorous mammal of the family Phocidae or Otariidae, having a sleek, torpedo-shaped body and limbs that are modified into paddlelike flippers.

Boron is a soft, brown, amorphous or crystalline, non-metallic element, extracted chiefly from kernite and borax, and used in flares, propellant mixtures, nuclear reactor control elements, abrasives, and hard metallic alloys. Atomic number 5, atomic weight 10.811, melting point 2 300°C, sublimation 2550°C, specific gravity (crystal) 2.34, valence 3.

(3) A definition can be expanded by explaining, or by providing additional information about key terms linked to the term being defined.

Example

Consubstantiation is the Lutheran (after Martin Luther) doctrine that the body and blood of Christ co-exist with the elements of bread and wine during the Eucharist, which is the Christian sacrament also called Communion.

(4) Expanded definitions can provide information on how to produce or experience the phenomenon being defined.

Example

The sound [f] is a voiceless, labio-dental fricative. It is formed by placing the lower lip lightly against the upper teeth, closing the vellum, and forcing the breath out through the spaces between the teeth or between the teeth and the upper lip.

Weisman, H M
Basic Technical Writing Merril 1988 136

☐ Concluding Remarks

Not all writers will provide you with paragraphs of definition in the course of introducing you to new information. Most textbooks, however, offer a glossary of important terms in an Appendix. Some writers provide definitions of key terms at the beginning of each chapter, or by way of footnotes. A comprehensive English dictionary is indispensable - good dictionaries do not only define 'ordinary' English, but also offer compact definitions of many scientific concepts. (In writing this chapter we made use of *The Heritage Illustrated Dictionary of the English Language,* which offers diagrams as well.) Your task as a reader is threefold:

(1) Find the definition of any term or concept which you do not understand (whether this is buried in the text, in a footnote, or in the glossary).

(2) Establish T, C and D. If these are not explicitly given, deduce them, or consult other sources for the missing information.

(3) Make sure that you fully understand the concept, will remember it, and are able to use it. It is a good idea to keep a personal glossary of terms, in which you record your own formal definitions of any new or problematic terms or concepts.

☐ Summary

Descriptions

> A paragraph of description describes physical appearance, function, or procedure.

(1) Physical description provides information on the physical characteristics of an object.

(2) Function description usually provides the reader with information about a device of some kind.

(3) Process description paragraphs describe a procedure, which may be natural or contrived.

> A good way to summarise a paragraph of physical description is to attempt to draw an accompanying diagram, using the information given in the text.

> Paragraphs containing function descriptions can be summarised by listing the events in a numbered sequence.

> Paragraphs describing a process can best be summarised by drawing a flow chart, or by listing a numbered sequence.

Definitions

> A definition is an explanation of the meaning of a word, phrase, concept, or phenomenon.

As a reader, your job is to be on the lookout for three basic 'chunks' of information, regardless of the format or type of definition presented by the writer. The crucial items are:

- ☐ the term [T] being defined
- ☐ the set, class or category [C] to which the object or phenomenon belongs, or of which the object is a subset
- ☐ the characteristics or differences [D] which distinguish the object or phenomenon from others in the same category, or which determine the subset.

☐ Bibliography

The Heritage Illustrated Dictionary of the English Language.
Morris, William (ed) McGraw-Hill
International Book Company. 1973.

☐ Answers

Practice session 1

(1) T : manticore
    C : fabulous monsters
    D : the head of a man
        body of a lion
        the tail of a dragon or scorpion

(2) T : homiletics
    C : fields of study
    D : dealing with the art of preaching

———— *continued* ————

continued

(3) T : scotopia
    C : abilities
    D : to see
        in dim light

(4) T : misogamy
    C : dislikes
    D : pathological
        of marriage

(5) T : zloty
    C : basic monetary unit
    D : of Poland

(6) T : xyster
    C : surgical instruments
    D : for scraping bones

(7) T : sphragistics
    C : fields of study
    D : of seals and signets

(8) T : pizza
    C : Italian baked dishes
    D : a shallow pielike crust
        covered usually with a spiced mixture of tomatoes and cheese

Practice session 2

(1) T : brake
    C : devices
    D : used for slowing or stopping motion
        usually by contact friction

(2) T : cold
    C : viral infections
    D : inflammation of the mucous membranes of the respiratory passages
        accompanying fever
        chills
        coughing
        sneezing

continued

—— *continued* ——

(3) T : pyrotechnist
    C : person
    D : designs fireworks
         manufactures fireworks
         sets off fireworks

(4) T : pencil
    C : implements
    D : narrow and generally cylindrical
         consisting of a thin rod of graphite or similar substance
         encased in wood, metal or plastic
         used for writing

(5) T : office
    C : places
    D : services
         clerical work
         professional duties are carried out

(6) T : diphthong
    C : speech sounds
    D : begins with one vowel sound
         moves to another vowel or semivowel
         within the same syllable

(7) T : interrobang
    C : punctuation marks
    D : combination
         question mark and an exclamation mark

(8) T : oncology
    C : fields of study
    D : deals with tumours

(9) T : nautical miles
    C : units of measurement
    D : based on the length of one minute of arc of a great circle
         equivalent to 1 852 metres

(10) T : succotash
    C : American dishes
    D : kernels of corn and lima beans cooked together

7

Types of Paragraphs (2)

> In the beginning God created the heavens and the earth.
> Now the earth was formless and empty, darkness was over
> the surface of the deep, and the Spirit of God was hover-
> ing over the waters.
> And God said, 'Let there be light,' and there was light.
> God saw that the light was good, and he separated the light
> from the darkness. God called the light 'day' and the dark-
> ness he called 'night'. And there was evening, and there
> was morning – the first day.
>
> Genesis 1, 1-5

☐ Classifications

This chapter continues the examination – begun in Chapter 6 –
of the types of information presented by paragraphs. The last chapter discussed paragraphs of description and definition, and looked at
strategies which you as a reader can use to tackle and integrate the
data presented by such paragraphs. In this chapter, we will examine
paragraphs of classification, with the aim of equipping you with aware-
ness and additional reading skills to deal with this type of discourse.

Definition

> Classification is the systematic grouping of items or phenomena
> into classes or categories on the basis of shared characteristics
> or traits.

Features of classifications

(1) The human brain has an innate desire for and capacity to per-
ceive pattern; and organising raw data so that it makes sense is
an instinctive urge. (It is significant that the first acts ascribed
to the Creator involve creating form out of formlessness, separating
out of different things, and naming of the resultant phenomena.)
Classifications, then, are an inevitable scientific response to the
apparant chaos of the observable world.

(2) A complete classification comprises the following information:
- ☐ the item(s) being classified; that is, the member(s) of the group or class
- ☐ the name of the class or group to which these items are assigned
- ☐ the basis or bases for classification; that is, the principle(s) according to which the items have been assigned to their respective groups or classes.

Example

There are many different kinds of musical instruments. They are divided into three main classes according to the way that they are played. For example, some instruments are played by blowing air into them. These are called *wind* instruments. In some of these the air is made to vibrate inside a wooden tube, and these are said to be of the *woodwind* family. Examples of woodwind instruments are the flute, the clarinet, and the bassoon. Other instruments are made of *brass*: the trumpet and the horn, for example. There are also various other wind instruments such as the mouth-organ and the bagpipes.

Some instruments are played by banging or striking them. One obvious example is the drum, of which there are various kinds. Instruments like these are called *percussion* instruments.

The last big group of musical instruments are the ones which have strings. There are two main kinds of *stringed* instrument: those in which the music is made by plucking the strings, and those where the player draws a bow across the strings. Examples of the former are the harp and the guitar. Examples of the latter are the violin and the cello.

Wallace, Michael J
Study Skills in English Cambridge University Press 1980 55

(3) Paragraphs of classification do not necessarily list every single member of every class. Instead, they provide an explication of the classifying principle, and sufficient examples of the principle in operation so that the reader can independently assign unallocated items to the appropriate class. (Note, for example, how the paragraph above explains the classifying principles, and offers only one or two examples of each group or class.)

Your task as reader, therefore, is primarily to ensure that you can identify, understand, and can use the basis for or principle of classification. Once the basis for classification is understood, it is easy to assign a new item to its correct place in the overall scheme or pattern of data.

Example

Class : percussion instruments
Basis for classification : played by banging or striking
Members : drum, etc.

(4) Classes of items can be further divided into sub-classes, which in turn can themselves be sub-divided, creating a hierachical classification. For example, in the paragraph on musical instruments, the class of stringed instruments is further divided into the class of stringed instruments played by plucking, and the class of those played with a bow. A different basis for classification operates at each point of division:

☐ these instruments are classified as musical instruments on the basis that this is their function;
☐ they are classified as a sub-group of the class of musical instruments on the basis of their producing sound from strings; and
☐ they are further subdivided into two groups on the basis of how the strings are manipulated to produce sound.

(5) Paragraphs of classification are best analysed and summarised by drawing up a classification tree, setting out each level of classification, and indicating the given members of the various classes.

Example

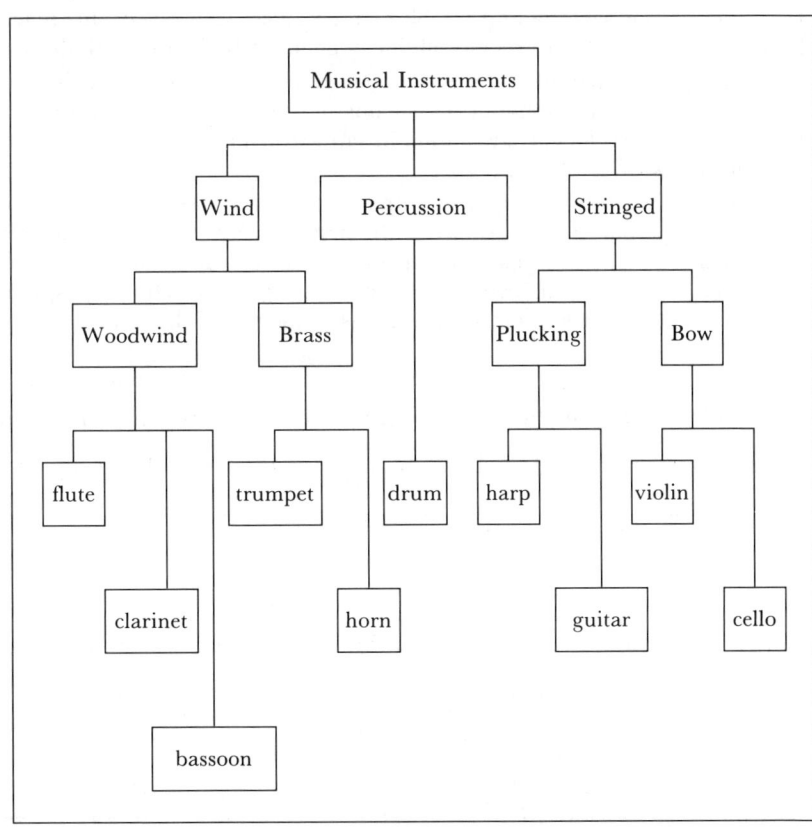

Practice session 1

Read the passage below, and then answer the questions that follow.

We start with the kingdoms, which for a long time were assumed to be two in number: animals and plants. ... However, the growing knowledge concerning the micro-organisms complicated matters, and the American biologist Robert Harding Whittaker suggested that living organisms be divided into no fewer than five kingdoms.

By Whittaker's system, the plant kingdom and the animal kingdom are confined to multicellular organisms. The plants are characterized by the possession of chlorophyll (so that they are the so-called green plants) and the use of photosynthesis. The animals ingest other organisms as food and have digestive systems.

A third kingdom, the fungi, are multicellular and resemble plants in some ways but lack chlorophyll. They live on other organisms though they do not ingest them as animals do, but excrete digestive enzymes, digest their food outside the body, then absorb it.

The remaining two kingdoms contain one-celled organisms. Protista, a word coined in 1866 by the German biologist Ernst Heinrich Haeckel, includes the eukaryotes: both those that are made of cells resembling those that constitute animals (protozoa, such as the amoeba and the paramecium); and those that are cells resembling those that constitute plants (algae).

Finally, a kingdom known as monera contain the one-celled organisms that are prokaryotes – the bacteria and the blue-green algae. Left out of this scheme are the viruses and viroids which are subcellular and might well form a sixth kingdom.

The plant kingdom, according to one system of classification, is divided into two main phyla – the Bryophyta (the various mosses) and the Tracheophyta (plants with systems of tubes for the circulation of sap), which includes all the species that we ordinarily think of as plants.

This last great phylum is made up of three main classes: the Filicineae, the Gymnospermae, and the Angiospermae. In the first class are the ferns, which reproduce by means of spores. The gymnosperms, forming seeds on the surface of the seed-bearing organs, include the various evergreen cone-bearing trees. The angiosperms, with the seeds enclosed in ovules, make up the vast majority of the familiar plants.

Asimov, Isaac
Asimov's New Guide to Science Penguin 1984 699

(1) Complete the classification tree, using the information given in the passage.

(2) State the basis for classification that differentiates fungi from animals.

....

....

....

(3) State the bases for classification that distinguish plants from animals.

....

....

....

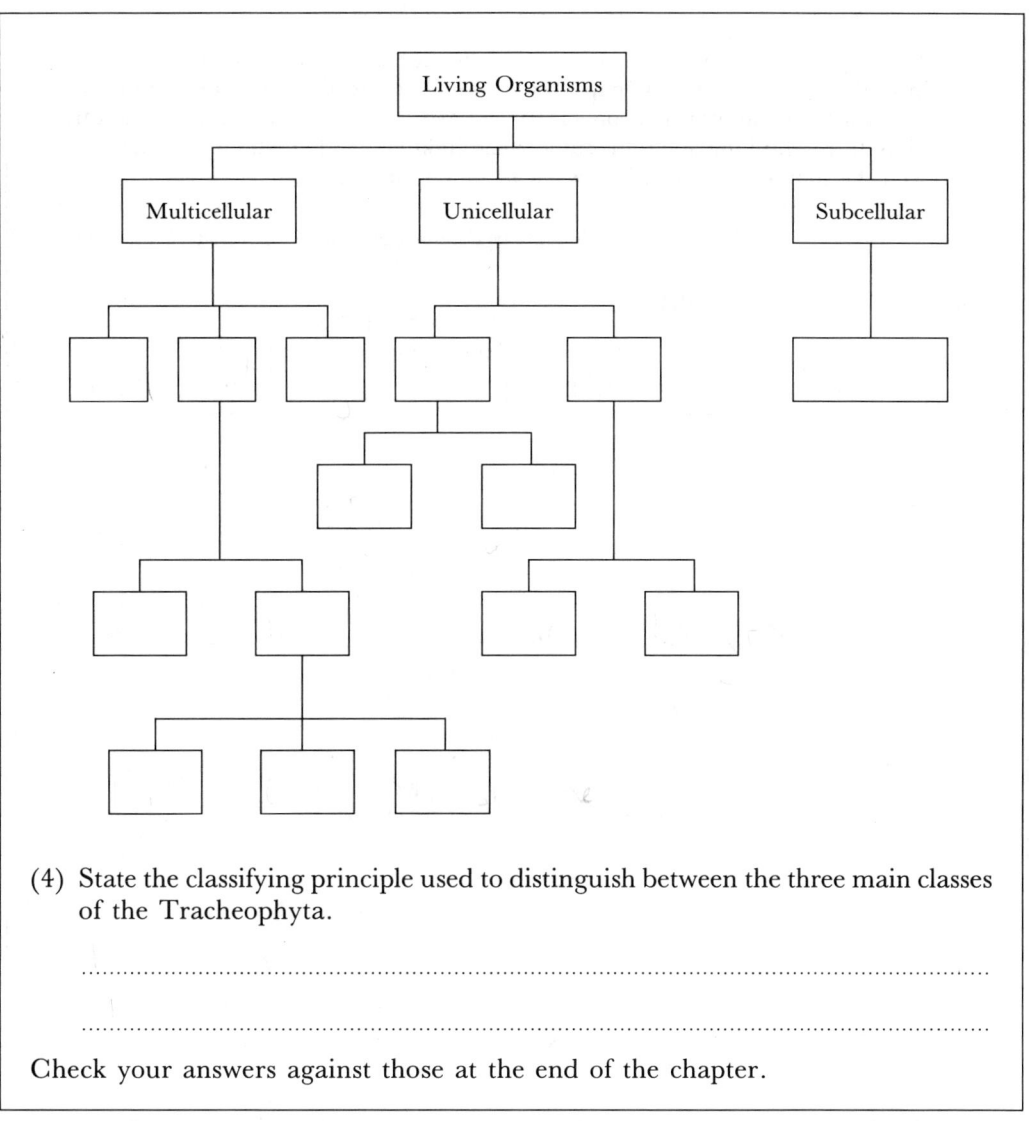

(4) State the classifying principle used to distinguish between the three main classes of the Tracheophyta.

....

....

Check your answers against those at the end of the chapter.

Practice session 2

Draw and complete classification trees for the following paragraphs.

(1) What are the ways in which a student can acquire information? Firstly, he will acquire information from his tutors, in three main ways – by lecture, by tutorial and by handouts which his tutor may give him. Secondly, he may acquire information from 'other experts' outside his college: principally by reading but also perhaps by listening to the radio, listening to cassette recordings, or watching educational

continued

continued

TV programmes. Thirdly, he will get information from his fellow students: perhaps in student-led seminars, perhaps in the contributions of other students in tutorial, or perhaps just in informal conversation. Lastly, he can acquire information from himself! By thinking about his subject and linking together what he has heard and seen, he may come up with new ideas, which are his alone.

Wallace, Michael J
Study Skills in English Cambridge University Press 1980 56

```
┌─────────────────────────┐
│  Sources of Information  │
└─────────────────────────┘
```

(2) The first baryons beyond the proton and the neutron to be discovered were given Greek names. There was the lambda particle, the sigma particle, and the xi particle. The first came in one variety, a neutral particle; the second in three varieties, positive, negative, and neutral; the third in two varieties, negative and neutral. Every one of these had an associated antiparticle, making a dozen particles altogether. All were exceedingly unstable; none could live for more than a hundredth of a microsecond or so; and some, such as the neutral sigma particle, broke down after a hundred trillionth of a microsecond.

Asimov, Isaac
Asimov's New Guide to Science Penguin 1984 328

continued

continued

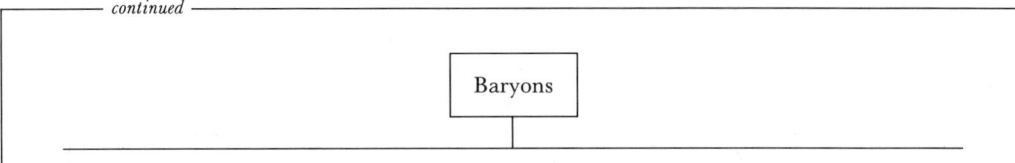

Baryons

(3) The writing process can be divided into five phases. Pre-writing is the stage prior to any actual need to produce a piece of written work. It can be seen as the ongoing process of nurturing and improving one's writing skills. Techniques used in the pre-writing stage include journal keeping and freewriting. The invention phase begins once the writer faces a specific writing task and has to 'invent' material to fill the page. Brainstorming, clustering, and heuristics are among the strategies applied in this process. drafting follows invention, and it is here that knowledge of paragraph patterns and essay structure is used. In the revision stage, nutshelling (reducing each paragraph to its main idea) and marking of signpost words are helpful techniques. The final stage is the editing of the manuscript, in which attention is paid to grammar, spelling, style, and layout.

continued

continued

```
                    ┌─────────────────────┐
                    │ The Writing Process │
                    └─────────────────────┘
```

Check your answers against those at the end of this chapter.

Practice session 3

Draw up a classification tree for the following data, presented to you in a jumbled form. Amongst the data are the names of major classes, sub-classes, and individual members of these. The top level of the tree is: Forms of Communication.

Human messengers

Non-electronic methods

Weather satellites

Using wires

Birds as messengers (pigeons)

Smoke signals

Semaphore

Horns (car, fog)

Cables

Signals

Radio-telephone

Telephones

Cinema

Use of satellites

Signals that can be seen

Messengers

Without using wires

Lighthouses

Radio

Electronic methods

Newspapers

Communication satellites

Handwriting

Drums

Signals that can be heard

Television

Printed books

Navigation satellites

continued

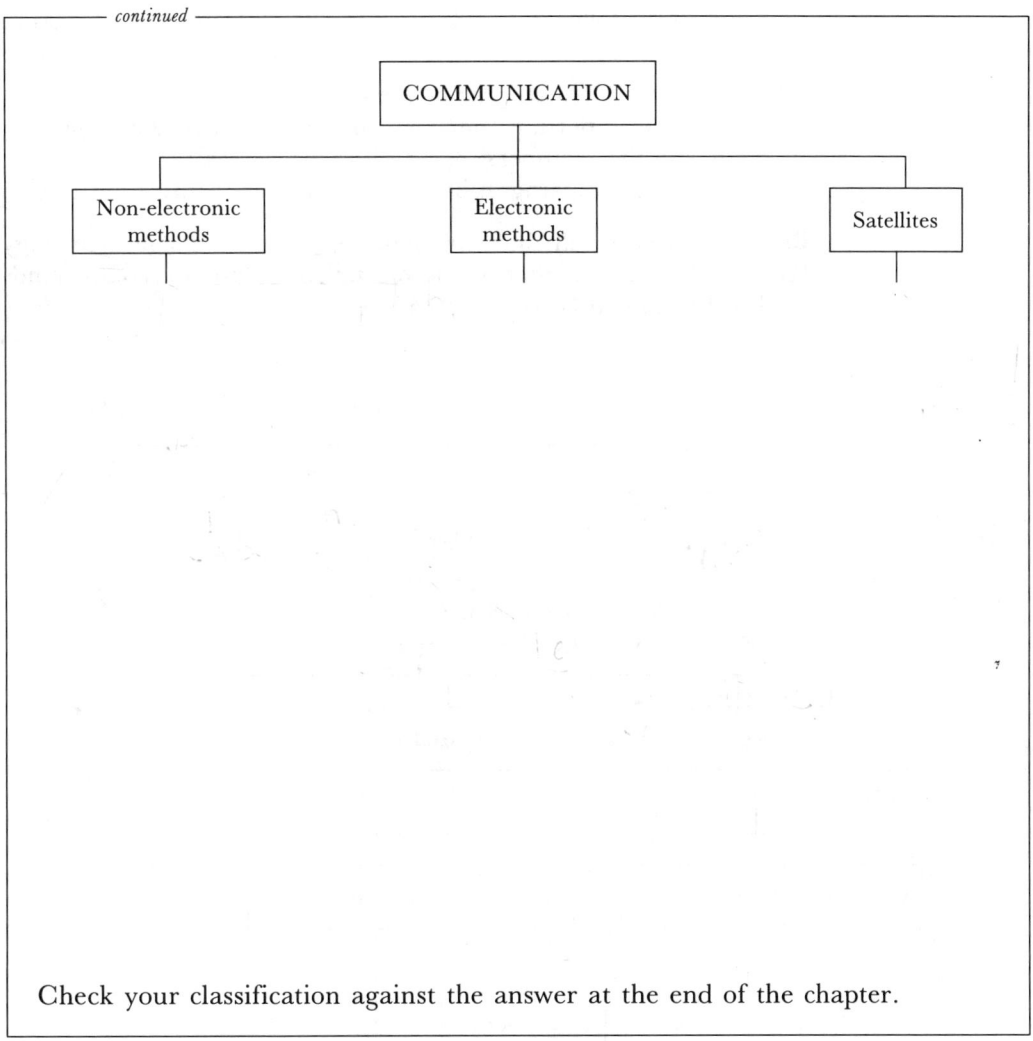

continued

Check your classification against the answer at the end of the chapter.

☐ Summary

Students of the sciences are often swamped with masses of data. Classifications are a way of making the data more manageable, but paragraphs of classification must be read with a strategy if they are to be of any assistance. Skimmed or read passively, these paragraphs will simply mean nothing to you.

Set yourself the following goals when reading a paragraph of classification:

☐ Identify the classifying principle(s).
(It is more important that you understand these than that you memorise the members of the different classes.)
☐ Summarise the paragraph(s) by drawing a classification tree.

Biology students will find the chapter on the language of biology (Chapter 9) useful in offering more detailed explanations of the kinds of classifications used in their science.

☐ Answers

(1)   **Practice Session 1**

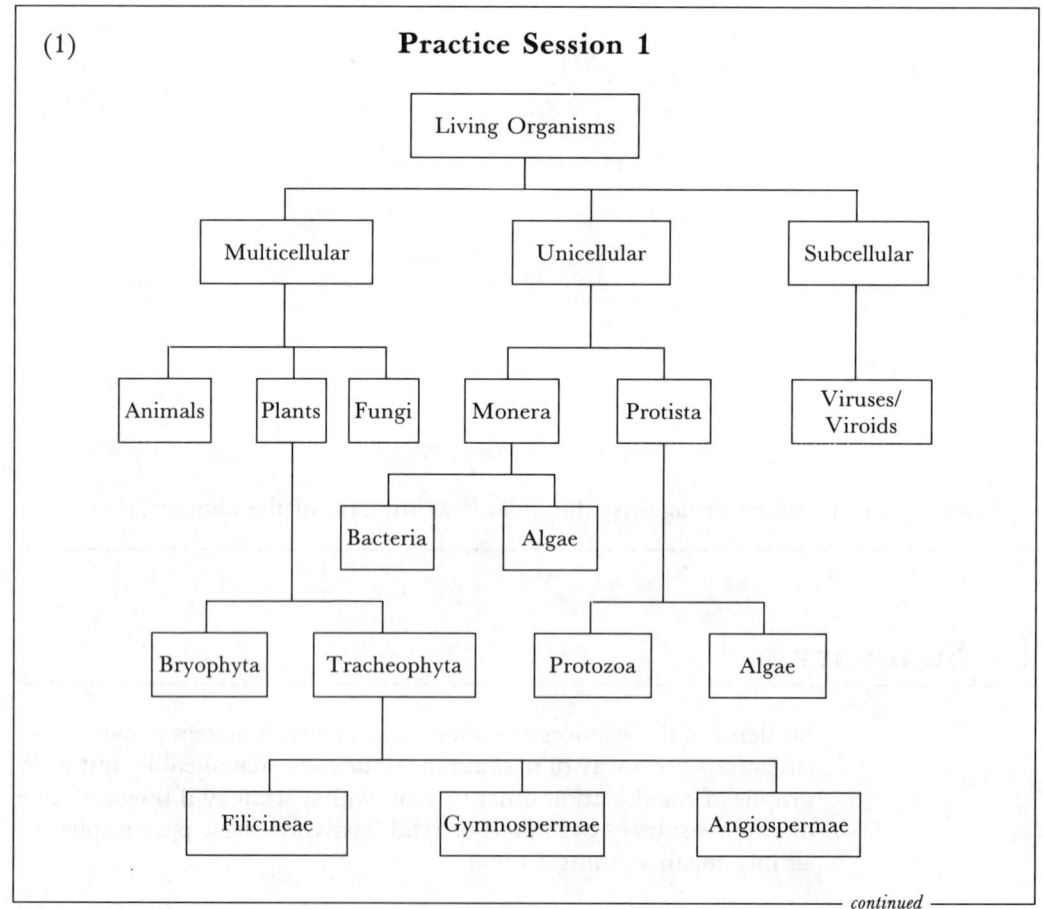

continued

continued

(2) Animals ingest other organisms in order to digest them, while fungi digest other organisms *outside* their bodies.

(3) Plants possess chlorophyll, animals do not.
Plants use photosynthesis to obtain energy, animals ingest other organisms.
Animals have digestive systems, plants do not.

(4) The Tracheophyta are divided into three classes according to their different methods of reproduction.

Practice session 2

(1)

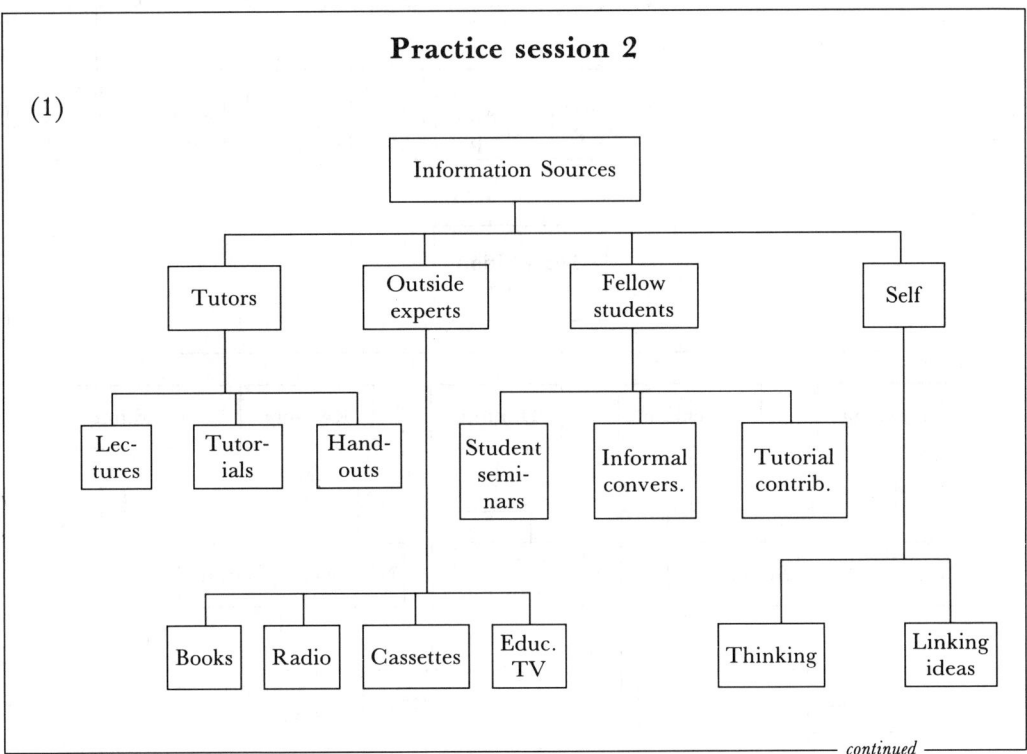

continued

continued

(2)

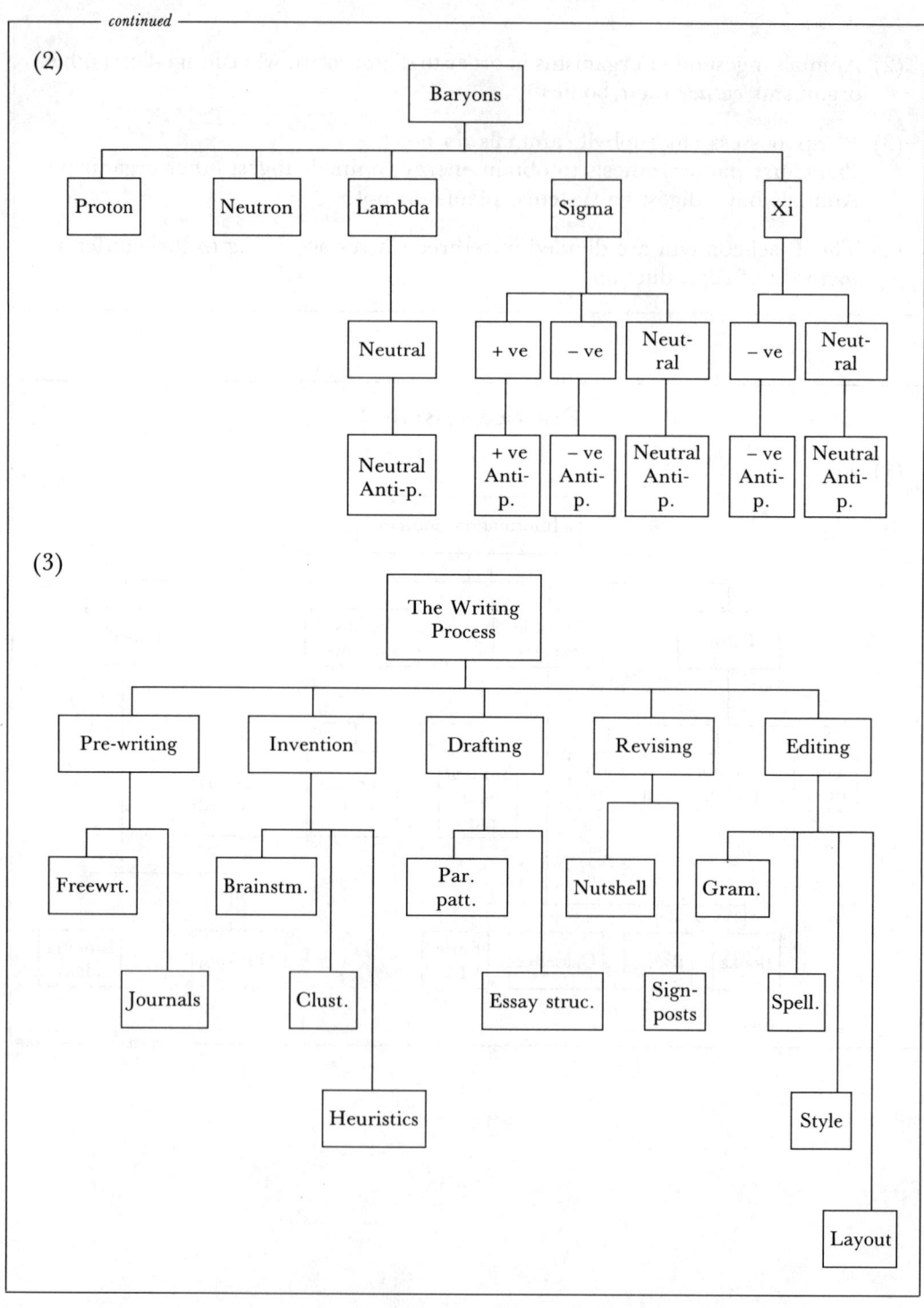

Practice session 3

The three major classes break down as follows:

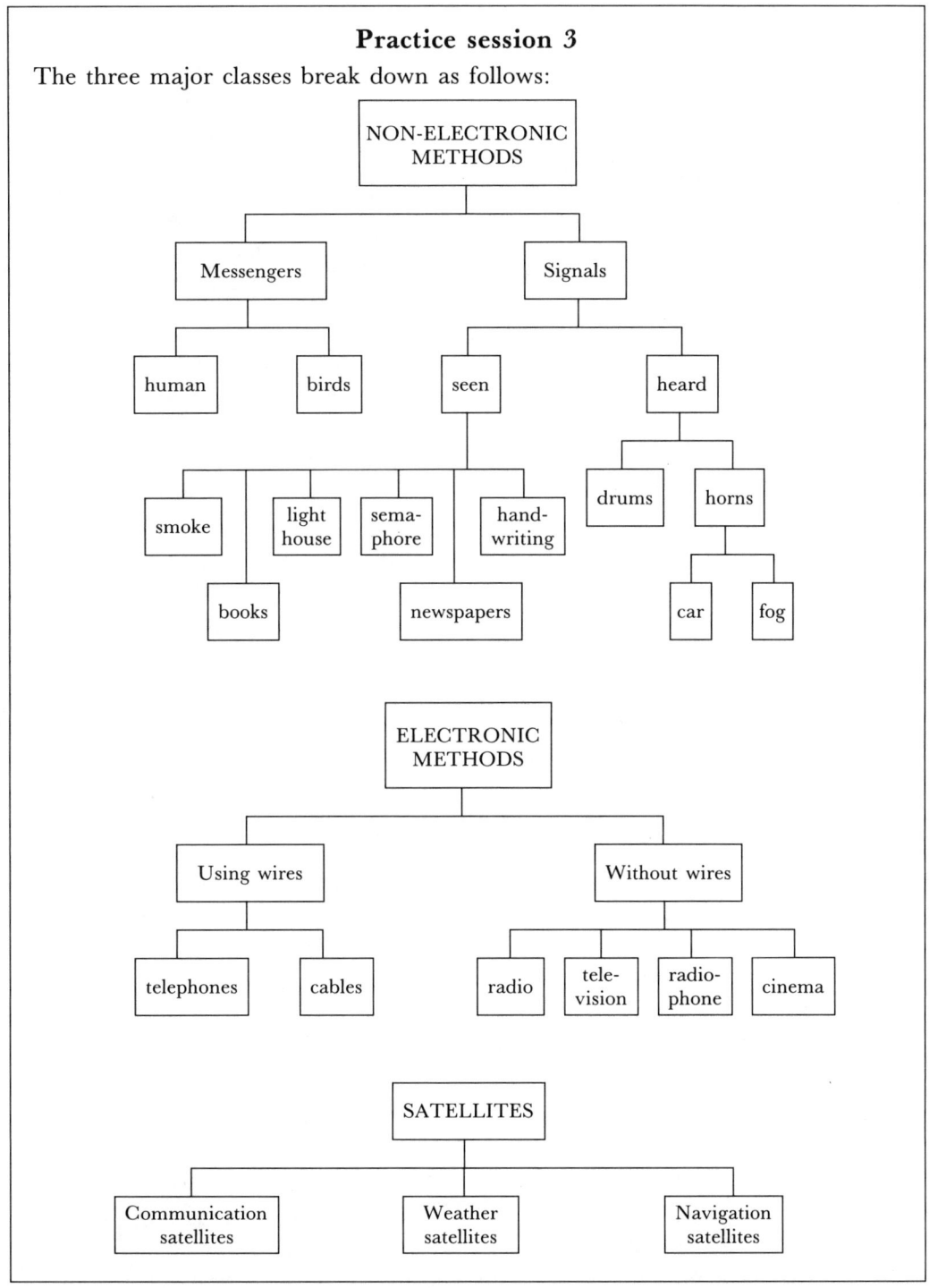

8

Logical Thinking

Brain: (n) an apparatus with which we think that we think.

Ambrose Bierce
The Devil's Dictionary

Chapters 1 to 7 dealt with with extracting information from text –
from words, from sentences, and from paragraphs. In this chapter
we are going to look at effective *use* of information – the skill called
logical thinking. The sciences demand a high degree of systematic
and intelligent manipulation and integration of quite complex masses
of data, and students need to develop a reactive, interactive, and creative reading and thinking style in order to progress successfully. This
chapter will offer you a number of strategies and suggestions for
empowering your reading and thinking processes, but these will be
effective only insofar as you invest time and energy in practising them.
Logic puzzles and problems (which will be used to teach the strategies in this chapter) can induce panic and avoidance reactions, and
practice sessions are often the part of a chapter that students automatically skip. Be warned! If you do not work through the exercises,
you will not find this chapter much help in making your thinking more
efficient.

(Solutions to the problems will be found at the end of this chapter.)

Definition

> Logical thinking is a sequence of mental actions aimed at organis-
> ing and manipulating information.

Many students believe that insight and understanding arrive in a blind-
ing flash of light. These students frequently feel frustrated when they
face a baffling problem or read – over and over again – a dense
explanation of a complicated concept, and wait in vain for that 'cos-
mic lightning'. They feel that other, more intelligent students seem
to understand and deal with complex information so much more intui-
tively and instantly – believing that they must be lucky enough to
have regular flashes of brilliant insight.

In fact, insight almost always follows an extended period of intensive grappling with the information, and hardly ever arrives fully-formed on first reading, or first sight of a problem. Effective thinkers do not rely on intuition but on systematic use of strategies. They are often unaware that they are doing this, because the processes are often unconscious and – as a result of frequent use – fairly rapid. Researchers studying effective thinkers have identified a number of tactics which both refine and 'rev-up' one's thinking skills. Basically, logical thought is getting information under control and then using it. The procedures, strategies, and tactics which this chapter describes are ways of helping you achieve this goal.

☐ Information Management

The first priority is to establish who's boss. As long as you allow yourself to be intimidated, overwhelmed, and swamped by masses of facts, your thinking abilities will be inhibited – cringing in some inaccessible corner of your brain, whimpering and useless. An essential start, which will also give you confidence, is to take the information by the scruff of its neck, and *organise* it.

---

(1) Always work with pencil and paper.

(2) Establish what is known, what is given, what the facts are. List them.

(3) Reduce ambiguity.
If something is confusingly phrased, re-state it in ways that make sense.

(4) Eliminate extraneous factors.
If there are facts or variables which are irrelevant to the central problem, cross them out.
Focus only on key variables and relationships.

(5) *Visualise* the problem.
Draw diagrams, graphs, charts – whatever will help you to get a more concrete grasp of the situation.

(6) Establish what the point of all the information is.
What are you supposed to do with the data; what are you required to solve?
Make sure that you know *why* you are wrestling with this information, and what the desired outcome is.

Pencil and paper

Use paper extravagantly. Thinking in written form is far more effective than bouncing ideas off the inside of your head. Your written notes do not have to be (and probably should not be) in flowing prose, or even in full sentences. You probably have your own form of short-hand (arrows, equal signs, etc), and will often only need to jot down key words, or draw diagrams. A desk surrounded by scrunched up bits of paper is one where serious thinking is happening!

Establish the facts

Do not stare at a mass of data in the hope that you will be able to ingest it whole.

Break any problem, theory, situation, concept down into its individual constituent variables.

List these facts or variables separately.

Example 1

A man is looking at a picture on the wall.

He says, 'Brothers and sisters have I none, but that man's father is my father's son.'

Who is in the picture?

What are the facts?

There is a man (Man 1)
There is a picture of a man ('that man') (Man 2)
Man 1 has no brothers and sisters
Man 2 has a father
That father is man no 1's father's son.

Reduce ambiguity

Facts may be wordily and confusingly stated. Try to reduce each variable to its simplest expression.

Example 1 (continued)

A man is looking at a picture on the wall.

He says, 'Brothers and sisters have I none, but that man's father is my father's son.'

The statement 'Brother's and sisters have I none' is confusingly worded.

Re-stated: Man 1 is saying, 'I have no brothers and sisters.'
Alternatively, 'I am an only child.'

Insert the rephrased variable into the problem.

A man is looking at a picture on the wall.
He says, *'I am an only child,*
but that man's father is my father's son.'

The statement 'my father's son' is wordy.

Re-stated: Since man 1 is an only child, his father's son can only be himself.

Insert the rephrased variable into the problem.

A man is looking at a picture on the wall.
He says, 'I am an only child,
but that man's father is *myself*.'

(Can you solve the problem now? Who is the man in the picture?)

Eliminate extraneous variables

All information is not equal. It is crucially important that you are able to distinguish between relevant, important facts, and facts that do not contribute anything significant to your understanding of the problem.

Zoom in on the key issues in a problem, and ignore all irrelevant data. For example, in the above problem, the fact that the man has no sisters is essentially useless information in that it doesn't add anything to our grasp of the issue, and is not a central concern, as the problem obviously involves men, fathers, and sons. (The answer, if you're still baffled, is that the picture the man is looking at is of his son.)

Irrelevant information usually rapidly reveals its worthlessness, because pursuing or investigating it usually leads one into a fruitless dead end. Learning to distinguish the important variables from the insignificant trivia will save you time and energy, and spare you much frustration.

Example 2

Four gangland hitmen, Faizal, Jabu, André and Nick, own between them 86 weapons made up of knives, handguns, and rifles. Faizal likes to be close to his work, so he owns 8 knives, 13 handguns, but no rifles. Jabu is just the opposite. He prefers to remain at a distance from violence and therefore owns only high-powered rifles, 33 in all. André is an all round professional who owns 5 knives, 8 rifles, and various handguns. Nick, also a man of many talents, has a total of 15 weapons, among which are 3 knives and 5 handguns. Interestingly, among them, the men own more than twice as many rifles as handguns, 48 rifles in all.

How many handguns do the men own among them?

The details about the men's personalities are irrelevant, and the wordy prose tends to obscure the facts. An essential first step in dealing with this problem is to eliminate all the excess information, and list only the relevant details.

Faizal – 8 knives 13 handguns
Jabu – 33 rifles
André – 5 knives, 8 rifles, ? handguns
Nick – 3 knives, 5 handguns, ? rifles = 15 weapons
Total rifles = 48 (more than twice number of handguns)
Total weapons = 86

Not all unnecessary information is this obvious, however. Sometimes you have to read carefully to identify key information as opposed to subsidiary details. Signpost words are often a good clue, marking out what is important versus what is secondary.

Example 3

Dracula hates daylight more than Wolfman unless Frankenstein hates daylight more than Dracula. In that case Wolfman hates daylight more than Dracula but less than Mummy. Mummy hates daylight more than Dracula but less than Frankenstein.
Arrange the four monsters in order from the strongest to the mildest hatred of daylight.

On first reading, this problem is discombobulating and panic-inducing, until one realises that the first two sentences are stating alternate possibilities dependant on a condition (whether or not Frankenstein hates daylight more than Dracula). The signpost words 'unless' and 'in that case' should alert you to this. Only the third sentence gives you a fact. If you start with the fact, and ignore the extraneous variables for the time being, you are well away to solving the problem.

Strongest hatred of daylight ◄─► Mildest hatred of daylight
Frankenstein ………… Mummy ………… Dracula

Now look at the first two sentences again and decide which of the the conditions apply.

Visualise   Drawing pictures to help you think logically has nothing to do with artistic ability. It merely involves putting the information down in a concrete, graphic form. In the section on paragraphs of description (Chapter 6) we noted how powerful diagrams can be in conveying information. Visualising a problem or concept often makes solving or understanding it a much simpler task.

Example 4

Four friends have always lived in the same town in four different residential areas: North Garden, South Garden, East Garden and West Garden, which lie, respectively, due north, due south, due east and due west of the town centre. From the clues that follow, can you determine in which residential section each of the four lives now and where each lived prior to the move described in clue 1?

(1) On 30 July of last year, Bobby moved to the area where Kathy was living and vice versa, with the result that Kathy now lives farther north than Bobby.

(2) Before Bobby and Kathy moved last July, Emma drove straight ahead at the town centre when visiting Bobby.

(3) At present, Luke lives east of Emma and must turn left at the town centre to visit Kathy.

Try to solve this problem in your head, and you'll soon run into problems! A diagram is an essential aid in coming to grips with the information.

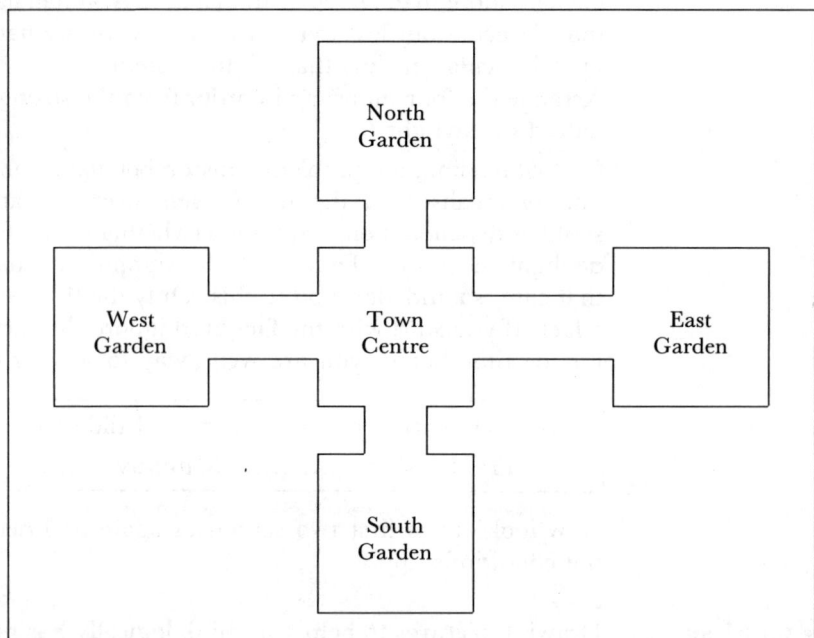

Example 2

The 'hitmen' puzzle on page 108 yielded the following information:

Faizal   – 8 knives 13 handguns
Jabu   – 33 rifles
André – 5 knives, 8 rifles, ? handguns
Nick   – 3 knives, 5 handguns, ? rifles = 15
Total rifles = 48 (more than twice number of handguns)
Total weapons = 86

This would be far more effectively presented in a table or chart.

	Knives	Handguns	Rifles	Total
Faizal	8	13	0	
Jabu	0	0	33	
André	5		8	
Nick	3	5		15
TOTAL			48	86

Setting out information graphically is often the step that leads to solving the problem, as the above example indicates.
(Solve this problem before continuing.)

Establish the desired outcome

Make sure that you know what you are required to do with the information. There is little point in manipulating information with dazzling dexterity only to arrive at a conclusion or answer that has no use or value. Before you start to process information, determine what your endpoint is to be.

Before starting on any problem, it is a good idea to visualise what form the answer is to take – a number, for example? what kind of number? more or less how big a number? Next to your list of facts (what you know), list what you don't know, and what you have to find out.

It is often useful to draw up an 'answer chart' with gaps indicating the information required.

Example 4 – Answer Chart

The solution for the problem of residential areas (North, South, East and West Garden) involves finding out the following information:

Name	Lived before 30 July	Lives now
Bobby		
Kathy		
Emma	XXXXXXXXXXXXXX	
Luke	XXXXXXXXXXXXXX	

(Try to solve this problem before continuing.)

Example 2 – Visualise the Answer

In the 'hitmen' puzzle, we are told that between them the men own more than twice as many rifles as handguns, 48 rifles in all, and you were asked to calculate the number of handguns the men own. You should have automatically formed an idea of what the answer was going to look like, more or less. Obviously, it is going to be a number; it is a number less than 24 (half of 48), but probably more than 19 (otherwise the number of rifles would be three times as many as the number of handguns).

Once you have solved the problem, compare your actual answer to the visualised answer. If there is a drastic discrepancy (for example, if your answer was 36 handguns), then you need to double check your calculations and the steps in your solution process.

☐ Information Manipulation

Sometimes, merely organising the information coherently is sufficient to solve the problem or provide you with the necessary insight. More often, however, once the information is organised you will need to manipulate or process it. Processing information is also not a magical art, but a skill which is dependent on systematic application of various strategies.

Strategies

(1) Break the problem down into separate tasks or steps.

(2) Systematically sequence and link tasks – conclusion of one leads on to the next.

(3) Consider implications of each fact.

(4) What if you reach a dead-end?
Try if-then reasoning.
Generate possibilities.

(5) Always double check against original premises.

Divide and rule

Successful thinking involves being able to think through a problem in discrete, separate stages. Do not try to engulf it whole and arrive at complete understanding instantly. Divide the problem into separate, smaller problems. Solve the bits that you can solve, and very often you will find that other parts of the problem become less daunting, or suddenly make sense.

Sequencing	Think in a sequence – as you arrive at one conclusion, be alert for the next logical step that the conclusion leads to, or the options that each conclusion generates.
Consider the implications of each fact	Do not accept all statements at given value. Consider what each fact implies – what extra, hidden conclusions each statement leads to, or is based on.
Dead-end tactics	If you come to a dead-end in your chain of reasoning, the chances are you may be following the wrong track. One option, therefore, is to throw away all your rough notes, and start from scratch. Preferably do this after an interval where you do not think about the problem at all. Go off and do something else, and come back to a fresh start later.

Another possibility is that you have reached the limit of the obvious deductions and conclusions, and now have to explore diverging and hidden possibilities. You need to generate multiple possibilities and explore the potential of each. You do this by a process of 'if-then' reasoning. Start by thinking 'What if ... ?', and then look at the consequences of your hypothesis.

(Examples of this type of reasoning are given in the Sample Problems at the end of this chapter.)

Be creative, be daring, take risks. The great thinkers were not afraid to postulate really bizarre theories (The earth is spherical? It revolves around the sun? Don't be ridiculous!), and while many of them were no doubt disproved after having been tested by rigorous reasoning, some proved to be surprisingly tenable.

Double check	Always check your solutions against the original premises or constraints. At the conclusion of each step in solving a problem, check that your solution does not violate any of the basic facts or principles. A key phrase is 'rigorous reasoning'. There is no room in serious, logical thinking for vague or woolly conclusions that are more or less right – particularly if you are to build further reasoning and conclusions on these.

Practice session

This practice session contains six problems of different kinds on which to practise the tactics and strategies outlined in this chapter. Some problems are set out for you in a colour-by-numbers fashion, while others are left to you to organise and solve. These problems are an essential part of this chapter – they offer you the

continued

—— *continued* ——

opportunity to put the strategies into practice, and to tone up your flabby thinking. The solutions are set out at the end of this chapter, but don't consult them until you have given each problem your best shot.

Problem 1

Pat, Lee, Chris and Kit are four friends. They seldom meet during the week because their various jobs – journalist, teacher, estate agent, hairdresser – kept them fully occupied. But they meet every Saturday evening for a game of cards.
Last Saturday, when they played, Pat's partner was the journalist, who was seated on Kit's left. Lee partnered the hairdresser. If Chris is neither the journalist nor the estate agent, what is each person's occupation?

Stage 1: Information Management

Chart the two variables (names and occupations) against each other.

Name	Journalist	Teacher	Estate Agent	Hairdresser
Pat				
Chris				
Lee				
Kit				

We are told that Chris is neither the journalist nor the estate agent. Register this information on the chart by eliminating these possibilities, using X's.

Name	Journalist	Teacher	Estate Agent	Hairdresser
Pat				
Chris	X		X	
Lee				
Kit				

—— *continued* ——

———— *continued* ————

Stage 2 – Information Manipulation

IF Pat's partner was the journalist, THEN Pat's occupation cannot be journalist.

IF the journalist was seated on Kit's left, THEN Kit cannot be the journalist.

By elimination, Lee must be the journalist, and therefore cannot be the teacher, estate agent, or hairdresser.

You can register this information on the chart as follows:

Name	Journalist	Teacher	Estate Agent	Hairdresser
Pat	✗			
Chris	✗		✗	
Lee	✓	✗	✗	✗
Kit	✗			

Now, the statement 'Pat's partner was the journalist' can be re-stated as: 'Pat's partner was Lee.'

IF 'Lee partnered the hairdresser', THEN Pat must be the hairdresser, and cannot be the estate agent or the teacher.

This information is registered on the chart as follows:

Name	Journalist	Teacher	Estate Agent	Hairdresser
Pat	✗	✗	✗	✓
Chris	✗		✗	✗
Lee	✓	✗	✗	✗
Kit	✗			✗

It is now clear from the chart that the only possible occupation for Chris is as teacher, which leaves estate agent as Kit's occupation.

———— *continued* ————

———— *continued* ————

Solution

Pat – hairdresser   Chris – teacher
Lee – journalist   Kit – estate agent

[Note a possible source of ambiguity in the names. The four names are androgynous – ie they could label men or women. Beware of automatic assumptions – gender was irrelevant to this problem, but you probably came to some (unfounded) conclusions about the sex of the participants.]

Problem 2

There are four members on a committee – the chairman, the vice-chairman, the secretary, and the treasurer. Their names, not necessarily in the same order, are Matthew, Mark, Luke, and John.

The chairman and the treasurer are cousins, but Mark and Matthew are not related to each other. The vice-chairman's wife is called Mary, and the secretary's fiancee is called Jane. The secretary is older than both Mark and Matthew. Luke has recently had a disagreement with the treasurer. Matthew and Luke are the only ones who are married.

Assign each man to his right place on the committee.

Use the chart below to record and process information, as was done for Problem 1.

Name	Chairman	Vice-chairman	Secretary	Treasurer
Matthew				
Mark				
Luke				
John				

Solution

Matthew is the ....

Mark is the ....

Luke is the ....

John is the ....

(Check your answers against the solution at the end of the chapter.)

———— *continued* ————

continued

Problem 3

Crate 1   Crate 2   Crate 3

APPLES	APPLES & ORANGES	ORANGES

Three crates of fruit were delivered to the farm stall on Monday. Owing to the heavy weekend the packers had had, each crate was incorrectly labelled. Assuming that you were allowed to take only one fruit from only one crate, how would you go about establishing what each crate contained so that you could label them correctly?

Stage 1 – Information Management
Facts:

(1) One crate contains apples.

(2) One crate contains apples and oranges.

(3) One crate contains oranges.

(4) All three crates are incorrectly labelled.

Constraint:
We may sample one fruit from one crate.

Goal:
Establish which crate to sample from so that we can deduce what each of the three crates contains.

(Irrelevant and extraneous variables are the farm stall, the day of the week, and the packers' hangovers.)

Stage 2 – Information Manipulation
IF all the labels are wrong, THEN ...
Deduction 1: Crate 1 must contain either Oranges or Apples & Oranges.
Deduction 2: Crate 2 must contain either Oranges or Apples.
Deduction 3: Crate 3 must contain either Apples or Apples & Oranges.

Deductions from the given information take us this far and no further. We now have to try hypothetical solutions, and test these solutions against the facts.

Complete the following if-then chains to arrive at a solution.

Hypothesis 1: What if we take a fruit from crate 1?

IF it is an apple, THEN according to deduction 1, ...

....

THEN Crate 3 can only contain ....

and Crate 2 must contain ....

continued

———— *continued* ————

But, IF the fruit you take out is an orange, THEN ....

....

....

Hypothesis 2: What if we take a fruit from crate 2?

IF it is an apple, THEN according to Deduction 2, ...

....

....

THEN crate 3 can only contain ....

and crate 1 must contain ....

But, IF the fruit is an orange, THEN ....

....

THEN crate 1 must contain ....

and crate 3 must contain ....

Solution

Take a sample fruit from crate ....

Problem 4

Each of five traders, whose names were Beer, Cork, Pepper, Wood, and Wool, sent a single consignment of goods to one of the others, none sending to and receiving from the same person. The commodities were beer, cork, pepper, wood, and wool. No commodity had the same name as its sender or receiver. Beer's consignment went to Wood. Wood sent pepper to the sender of cork. Cork received beer, and the sender of wood had the name of the commodity received by Pepper.

What commodity did each of the five traders send, and to whom?

Stage 1 – Information Management

Five Traders: Beer   Five commodities: beer
             Cork   cork
             Pepper   pepper
             Wood   wood
             Wool   wool

Constraints:

(1) No commodity had the same name as its sender or receiver.

(2) No one sent to and received from the same person.

———— *continued* ————

———— continued ————

Known facts:

(1) Beer sent A to Wood.

(2) Wood sent pepper to B who sent cork to C.

(3) D sent beer to Cork.

(4) X sent wood – Pepper received x.

One way of tabulating the information is the following:

Sender	Commodity	Receiver
Beer	~~beer~~/~~cork~~/~~wood~~/(wool)	Wood
Cork	beer/cork/wood/wool	Beer/Pepper/Wool
Pepper	beer/cork/wood/wood	Beer/Cork/Wool
Wood	pepper	~~Beer~~/~~Cork~~/~~Pepper~~/(Wool)
Wool	beer/cork/wood/wool	Beer/Cork/Pepper

Stage 2 – Information Manipulation

Fact 1: Beer sent A to Wood.
Fact 2: Wood sent pepper to B who sent cork to C.

Deductions:

In terms of Constraint 1, A cannot be beer or wood.

In terms of Constraints 1 and 2, B cannot be Pepper, Cork, Wood or Beer.

Therefore, A cannot be cork (because B sent cork, and B is not Beer).

IF Wood sent pepper, THEN A cannot be pepper.

Eliminate these possibilities on your chart.

Sender	Commodity	Receiver
Beer	~~beer~~/~~cork~~/~~wood~~/ wool	Wood
Cork	beer/cork/wood/wool	Beer/Pepper/Wool
Pepper	beer/cork/wood/wood	Beer/Cork/Wool
Wood	pepper	~~Beer~~/~~Cork~~/~~Pepper~~/ Wool
Wool	beer/cork/wood/wool	Beer/Cork/Pepper

———— continued ————

———— continued ————

It is now clear that:

☐ Beer sent wool to Wood.
☐ Wood sent pepper to Wool.

Therefore:

☐ none of the other traders could have sent wool; and
☐ none of the other traders could have sent to Wool.

(Eliminate these possibilities on the chart as we did for the previous deductions).

Fact 2 now reads:

☐ Wood sent pepper to Wool who sent cork to C.
(Eliminate the other commodities next to Wool on the chart.)

Therefore:

☐ none of the other traders could have sent cork
☐ C cannot be Cork.

(Eliminate these possibilities on the chart.)

In terms of Constraint 1:

Cork could not have sent ....

Pepper could not have sent ....

Wool could not have sent ....

(Eliminate these possibilities on the chart.)

Fact 3: D sent beer to Cork.

In terms of Constraint 1, D cannot be ....

If D sent beer to Cork, Cork cannot have sent ....

(Eliminate this possibility on the chart.)

Therefore Cork must have sent ....

The only trader left who could have sent beer is ....

Therefore, D sent beer to cork means that ....
sent beer to Cork.

(Register this on the chart.)

Fact 4: X sent wood – Pepper received x.

Given the above deductions, X must be ....

Therefore Pepper received ....

Who sent this commodity? ....

Therefore, Cork sent to .... (the only
unassigned recipient).

———— continued ————

continued

(Register this on the chart.)

Solution

Beer sent .... to ....

Cork sent... to ....

Pepper sent .... to ....

Wood sent... to ....

Wool sent... to ....

(Double check your answer against the facts and the constraints, and then consult the answers at the end of the chapter.)

Problem 5

A boatman has to ferry three things across a river – a lion, a buck, and a large sack of cabbages. However, he can only fit one of these in his boat at a time. The lion will eat the buck if they are left alone together, and the buck will eat the cabbages given half a chance. How is he to ferry all three across the river intact?

Stage 1 – Information Management

Three variables – lion
                           buck
                           cabbages

Constraints:

(1) Only one can fit in the boat at a time.

(2) Lion + buck is an impermissible combination.

(3) Buck + cabbages is an impermissible combination.

Goal: Get all three items moved to the other side of the river.

Stage 2 : Information Manipulation

IF: Trip 1 – Takes the buck across.
    Leaves the lion and the cabbages behind.
    Returns, leaving the buck alone on the other side.

THEN: Option 1 – Take the lion.
      Option 2 – Take the cabbages.

IF he takes the lion and comes back for the cabbages (Option 1), THEN the lion will eat the buck.

IF he takes the cabbages and comes back for the lion (Option 2), THEN the buck will eat the cabbages.

continued

continued

What is the Solution?

....

....

....

....

....

(Compare your solution to that at the end of this chapter.)

Problem 6

Four talent scouts brought in five football players – a German, an Italian, a Rumanian, a Spaniard, and a Hungarian – and, having labelled each one with a different letter, they made the following statements to the team manager. Unfortunately, each talent scout made only ONE true statement:

Fibkin: A is not Spanish. B is not German. C is Italian.

Notzo: E is not Rumanian. A is German. D is not Italian.

Leiski: D is Spanish. B is not Rumanian. C is not Hungarian.

Truthizov: B is Hungarian. E is not Italian. C is not German.

What is the nationality of each of the five football players?

Be very careful to check your solutions against the original constraints – namely, each talent scout made two false statements and one true statement.

This problem is the acid test of your patience and discipline in logical thinking, so:

☐ approach it systematically, using pencil and paper;

☐ use if-then reasoning;

☐ take a break and start again if you get bogged down;

☐ don't give up and look at the answer at the end of this chapter too soon! (If you really get frustrated and stuck, read only the first sentence of the solution, and then see if you can solve the rest of the problem on your own.)

Solution

A is ....

B is ....

C is ....

D is ....

E is ....

☐ Answers and Solutions

Example 2

Name	Knives	Handguns	Rifles	Total
Faizal	8	13	0	21
Jabu	0	0	33	33
André	5	4	8	17
Nick	3	5	7	15
TOTAL	16	22	48	86

Example 3

Strongest hatred of daylight ◄————► Mildest hatred of daylight

Frankenstein .... Mummy .... Wolfman .... Dracula

Example 4

Name	Lived before 30 July	Lives now
Bobby	East Garden	South Garden
Kathy	South Garden	East Garden
Emma	XXXXXXXXXXXXXX	West Garden
Luke	XXXXXXXXXXXXXX	North Garden

Practice session

Problem 2

John is the secretary.

Luke is the chairman.

Matthew is the vice-chairman.

Mark is the treasurer.

— *continued* —

—————— *continued* ——————

Problem 3

Crate 1	Crate 2	Crate 3
APPLES	APPLES & ORANGES	ORANGES

Hypothesis 1: A sample fruit from crate 1 will enable us to re-label all the crates.

IF it is an apple, THEN according to deduction 1, crate 1 must contain Apples & Oranges. THEN Crate 3 can only contain Apples, and Crate 2 must contain Oranges.

But, IF the fruit you take out is an orange, THEN you do not know if the crate contains only oranges, or whether it contains apples as well.

Discard the hypothesis.

Hypothesis 2: A fruit from crate 2 will enable us to re-label all the crates.

IF it is an apple, THEN according to Deduction 2, crate 2 must contain Apples. THEN crate 3 can only contain Apples & Oranges, and crate 1 must contain Oranges.

But, IF the fruit is an orange, THEN crate 2 must contain Oranges. THEN crate 1 must contain Apples & Oranges, and crate 3 must contain Apples.

Solution

Take a sample fruit from crate 2.

Problem 4

In terms of Constraint 1:
Cork could not have sent cork.
Pepper could not have sent pepper.
Wool could not have sent wool.

Fact 3: D sent beer to Cork.

In terms of Constraint 1, D cannot be Beer.
If D sent beer to Cork, Cork cannot have sent beer.
Therefore Cork must have sent wood.
The only trader left who could have sent beer is Pepper.
Therefore, D sent beer to cork means that Pepper sent beer to Cork.

Fact 4: X sent wood - Pepper received x.

Given the above deductions, X must be Cork.
Therefore Pepper received cork.
Who sent this commodity? Wool.
Therefore, Cork sent to Beer (the only unassigned recipient).

—————— *continued* ——————

—— *continued* ——

Solution

Beer sent wool to Wood.
Cork sent wood to Beer.
Pepper sent beer to Cork.
Wood sent pepper to Wool.
Wool sent cork to Pepper.

Problem 5

Trip 1: Take the buck, leave it on the other side, return.
Trip 2: Take the lion, leave it on the other side, pick up the buck, return with the buck.
Trip 3: Drop the buck, pick up the cabbages, take the cabbages across, leave them on the other side with the lion, return.
Trip 4: Pick up the buck, cross to the other side.

All three items are now on the other side of the river.

Problem 6

Fibkin's second statement is true.
Notzo's first statement is true.
Leiski's third statement is true.
Truthizov's second statement is true.

A is Spanish
B is Rumanian
C is German
D is Italian
E is Hungarian

9

An Introduction to the Language of Biology

Saepe enim et verba non latina dico ut vos intelligatis.
Often, indeed, I use non-Latin words so that you may understand.

St. Augustine of Hippo (AD 354 - 430)
Commentary on Psalm 133

☐ Introduction

Biology (Greek *bios*: life; Greek *logos*: speech, discourse, reason, word) is the name of the science which encompasses the study of all living creatures, plants and viruses. Biology is subdivided into various branches, for instance:

☐ **botany** (Greek *botane*: grass, fodder) the science of plants
☐ **biochemistry** the chemistry of living organisms
☐ **biophysics** the physics of living organisms
☐ **cytology** (Greek *kytos*: a vessel, hollow) the study of living cells
☐ **microbiology** (Greek *micros*: little) the study of the microbes which infect man and animals
☐ **neurology** (Greek *neuron*: nerve) the study of the nerves
☐ **paleobotany** or **palaeobotany**[1] (Greek *palaios*: old) the study of fossil plants
☐ **palaeichthyology** (Greek *palaios*: old; Greek *ichthys*: fish) the study of fossil fishes
☐ **zoology** (Greek *zoion*: animal) the science of animals.

and very many more. For the purposes of our study of the language of science we will not concentrate on the language of each individual branch of the biological sciences, but will treat them all under the one heading of *biology,* since they share many of the words and technical expressions.

---

1 The prefix can be either *palae-*, *palaeo-* or *paleo-*.

127

Names   The biologist needs a vocabulary in order to communicate his discoveries and thoughts to his fellow scientists. The point of departure of all scientific communication is always the process of giving a name to something in such a fashion that everybody understands it. Names are given to:

- ☐ the living organisms existing now on earth in such a way that a universal picture of their relationships evolves
- ☐ the new life forms created by interbreeding programs and by (the slower) natural processes
- ☐ the life forms which previously existed on earth, the evidence of which is found in the fossil records (Latin adjective *fossile*: dug out) as preserved in the rocks
- ☐ the organs of all organisms (we call this branch of the science *anatomy*, from the Greek *ana*: up, and *tomos*: to cut) – from the smallest (the viruses) to the largest (the whale at present, and the huge *Sauria* or reptiles of previous aeons)
- ☐ the habits and living conditions of all living organisms (we call the study of this aspect of nature *ecology*, from the Greek *oikos*: house, and *logos*: discourse, speech, reason), their methods of reproduction and their growth patterns
- ☐ the life processes of all organisms on a macro scale, as well as on a micro scale (biochemistry), including the phenomenon we call sickness
- ☐ the actions and experiments we undertake to probe the phenomenon we call life, for instance, the analyses and instruments we use (such as a spectrometer, a microtome, a microscope, etc.).

These names can never be given in isolation, since biology is enriched by its contact with other sciences (such as chemistry and physics), and *vice versa*.

Biology thus has a vocabulary and a language all of its own[2]; it is superimposed on both the language of ordinary literary English, as well as on that of science in general as described in the first part of this book. Much of the nomenclature of biology is based upon *Latin*, a 'dead' language which was used as the language of communication

---

2   There are many dictionaries of biology in which technical terms are defined. One of the best is M Abercrombie, C J Hickman, *The penguin dictionary of biology*, Sixth Edition, Penguin Books, Hammondsworth, Middlesex, 1978 (or any later edition); no etymologies are given. Another general students' dictionary is E.A. Martin, *A dictionary of life sciences*, Pan Books in association with the Macmillan Press, London, 1978 (contains no etymologies). There are also dictionaries of most of the branches of biology which are listed above.

between scientists up to the sixteenth century, and in some cases, up to the eighteenth century. *Biological Latin* is best described as:

> a modern Romance language of special technical application, derived from Renaissance Latin with much plundering of ancient Greek[3].

Many of the words of *Biological Latin* appear in an almost unchanged and easily recognisable form in the various modern languages (like English) which are used by scientists to communicate in writing and in speech. Some of these words are also in common (sometimes very common!) use and we hear and read them daily without ever giving a thought to their origin. When we discuss the words of *Biological Latin* we are thus also including similar words in English at the same time. One of the main problems which a beginner faces is the pronunciation of the technical (and even of the common) biological words and technical terms. Knowing how these words are put together, makes it easier to pronounce them in a more or less correct or acceptable way.

Biological Latin is still the official language used by the *International Code of Botanical Nomenclature*[4] for the description of plants (for instance, more than 250 000 plants have now been described).[5] The international rules state that the name of a newly-discovered plant or animal must be accompanied by a *diagnosis* (Greek *dia*: refers to the concept of thoroughly, completely; Greek *gnosis*: knowing; the word thus means complete knowledge about something[6]), that is, by a statement of the characteristics of the plant.

The same holds for Zoology, where use is made of the *International Code of Zoological Nomenclature*[7] which is revised on a regular basis, and the annual *Zoological Record* of the Linnaean Society of London.

---

3 W T Stearn, *Botanical Latin*, Third revised edition, David and Charles, London, 1985.

4 International Association for Plant Taxonomy, *International code of botanical nomenclature,* (Adopted by the Fourteenth International Botanical Congress, Berlin, July-August 1987), Koeltz Scientific Books, Koenigstein, Germany, 1988.

5 Botanists are now engaged in an internationally co-ordinated project, based in England at the famed Kew Gardens. This project intends to describe all the plants in the world.

6 This is also the origin of the medical term *diagnosis* – the conclusion which we all expect the doctor to reach when he visits a sick patient. The complete knowledge of the problems of the patient gives the doctor the opportunity to make a judgement about the medicine to be prescribed.

7 International Union of Biological Sciences (IUBS), *International code of zoological nomenclature,* Third Edition, International Trust for Zoological Nomenclature in association with the British Museum (Natural History) London/University of California Press, Berkeley and Los Angeles, 1985. The *Code* also contains the full grammatical rules of Zoological Latin, rules for transliterating Greek into Latin, as well as a very good glossary in which all the relevant words needed to understand the book, are defined. The *Code* does not use the language of mathematical set theory which is used in the present book, but prefers the words of lower rank or of higher rank for the concepts superset and subset, respectively.

The Latin of Biology, although derived from Classical Latin, Medieval Latin and Renaissance Latin, is much simpler, for it is not used to express any deep thoughts and feelings or to write a poem or play. It is used for the factual (and rather 'dry' and accurate) description of the organisms, for nomenclature purposes and for naming and for systematisation. *Biological Latin* is thus an internationally accepted language:

☐ which forms part of every other modern language used by biologists
☐ which is used for the formal description of living organisms in the biological sciences
☐ which is specially designed for easy use, reference and consultation
☐ which is relatively easily understood by non-Latinists (provided they make some effort to learn how the biological Latin words are constructed from Latin and Greek roots and provided they make an effort to learn a basic vocabulary of words, prefixes and suffixes which appear over and over again).

Even though the structure of *Biological Latin* is mostly based upon Latin, there are many words which have classical Greek origins, or which are partly made up from Greek prefixes and/or suffixes which are added to the stem word[8]. Many such examples will be discussed in the text below; some of them already feature in the first few lines of this chapter.

Although the mainstay of English biological nomenclature and terminology is Latin and Greek, there are a number of terms which originated in some of the modern languages, like English, German, Dutch, Spanish, Italian, and many more. The following are some examples:

☐ *agar* or *agar-agar*: (Malayan) *noun,* a jelly made from the cooking of (red) seaweeds which is used in science for the culture of microorganisms, for making glue, for cooking, for giving stability to ice cream, etc.

☐ *Ginkgophyta*: (Japanese *ginkgo*: silver apricot plant; Greek *phyton*: plant), *noun,* a *Phylum* of which only one species (which is called the living fossil) survived from the Mesozoic era (65 – 248 million years ago), namely the maidenhair tree which has been cultivated in temples and other sacred sites in Asia for centuries; the trees are *dioecious*[9] (Greek: *di*: twice; Greek *oikos*: house), that is the sexes are on separate plants, while most of the other plants

---

8 The international organisations discourage the mixing of Greek and Latin in one word, but some older scientific words exist where the two languages are used together.
9 The word is pronounced something like *di-e-shesh*.

are *monoecious* (Greek *monos*: single), that is both sexes occur on one and the same plant;

☐ *hook*: the animal *Phylum* (Greek *phylon*[10]: race) called the *Pentastomata* (Greek *penta*: five; Greek *stoma*: mouth), also called the tongue worms because of their shape, are parasitic inhabitants of the air channels of dogs, and attach the *hooks* in their mouths to their hosts;

☐ *tun*: the animal *Phylum* called the *Tardigrada* (Latin *tardus*: slow; Latin *gradus*: step), the slow-steppers (which were called the water bears by Thomas Huxley in the nineteenth century) is a very resilient group, since they can survive temperatures in hot springs (up to about 151°C) and can even live after being cooled down to − 268°C, which is just above the absolute zero of temperature; they can roll themselves into roundish objects called *tuns,* from the Old English word *tunne*: wine barrel; these *tuns* can survive a century under very adverse conditions; the picture of one of these slow-steppers which is reproduced in **Figure 9.1**, illustrates the name and the nicknames of this animal.

Some of these words are only used in English biological texts; when a Latin description is required, then the proper Latin term is used. They are included here to illustrate the fact that not all English biological terms are derived from Latin. Such English and even Anglicised Latin and Greek words are used in school and university text books of biology. A good dictionary must always be available to look up the meanings. The *Chambers's Twentieth Century Dictionary* (W & R Chambers Ltd, Edinburgh and London, 1952; there are later editions available) and *The Concise Oxford Dictionary of Current English* (Oxford University Press, Sixth Edition, 1976) proved to be very valuable and assisted us to give descriptions of many word examples in this chapter. Dictionaries such as these are indispensable for studying.

In the sections below the etymology of many technical terms is discussed and their roots in Latin and/or Greek are identified. Many examples are cited to illustrate the principles governing the different ways of constructing such technical terms. These examples have been selected with great care so that they also:

☐ illustrate the most important of the original classical words which are used to construct technical terms in general

---

10 Some dictionaries transliterate the Greek word as *Phulon*.

☐ make it easy for the non-classicist to learn this core of classical Latin and Greek verbs, nouns, adverbs, prepositions, prefixes and suffixes which are used in scientific terminology.

Figure 9.1
Echiniscus blumi
Reproduced with permission from L Margulis and K V Schwartz, *Five kingdoms – An illustrated guide to the phyla of life on earth,* Second Edition, W H Freeman and Company, New York, 1988.

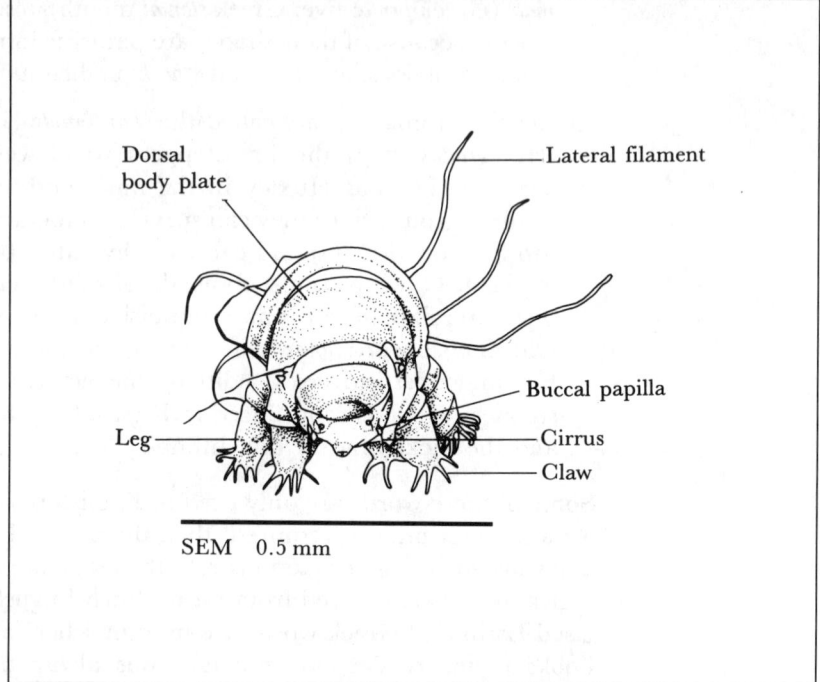

Echiniscus blumi (Greek *echinos*: a hedge-hog; surname of a person: *Bloom*, or *Blum*), an example of the *Phylum Tardigrada,* illustrating the ponderosity of the organism and its nickname water bear; a little imagination will show that it can easily transform itself into a tun by rolling up. The adjective *buccal* is derived from the Latin *bucca*: cheek; the noun *papilla* comes from the Latin diminutive of *papula*: a pustule, a pimple; the noun *cirrus* comes from the Latin *cirrus*: a curl or a tuft. Every name in a biological drawing should be looked up in a dictionary, or its meaning derived by means of the rules and the examples given below.

The abbreviation SEM stands for Scanning Electron Microscope. The solid black bar on the bottom of the picture represents 0.5 mm; it allows the size of the organism to be determined.

☐ The Language of Systematisation – Preliminary Discussion

The first human act which is described in the Bible is that of Adam giving names to the various animals which the Lord had brought to him to be named (*Genesis* 2:19); a charming picture of these animals which comes from the Luther Bible of 1736 is shown in **Figure 9.2**. We have kept this good habit which Adam started, and we are still giving names in all languages to living (and also to inanimate) entities and their component organs or parts. Let us think a bit about what we are actually doing when we give a name to a living thing (or to any one of its organs or smaller components, for that matter).

Figure 9.2
The animals being brought to Adam to be named – as depicted in the 1736 edition of the Luther Bible. This curious engraving even shows at least one animal pair which comes straight from pagan mythology! Can you identify them?

The naming of both animate and inanimate *macroscopic objects* (in contrast to objects at atomic and molecular level) depends upon a set of unspoken assumptions which we make, namely that

☐ all objects in the physical world are *individual entities,* each one having a unique set of properties or characteristics which can be observed and listed to any required degree of completeness (the more properties this list contains, the more precisely the object is described)

☐ the *complete set of properties* of an object distinguishes it absolutely from all other objects, that is, the listed properties individualise or personalise it

☐ this complete or full set of properties can be given a name, called a *noun,* which is used in speech and in other methods of communication in order to refer to that particular physical object[11]

☐ *subsets of properties* of all objects can be found which can be used to classify similar objects into categories or groups, each of which then refers to a set of similar objects or organisms

– two or more objects are considered to belong to the same category or group if the chosen subsets of their properties are identical

– such a subset can also be given a name, called a *noun,* which is used to denote the whole category or group to which the two or more similar objects belong[12]

– consecutive subsets of subsets of properties can be found by selectively deleting properties from a complete set so that classification schemes can be constructed.

As an example, let us consider my dog *Lucie* and the sequence of names or nouns which can be used to refer to her, as well as the full set of properties or characteristics which I discern in her, such as the following:

is alive, can move around, does not contain chlorophyll, is big, eats meat, has four legs, has canine teeth, has a skin with short tan and white hairs, has two eyes, has two ears, has bones, has feet, has nails on the toes, has pads under the feet, has a skeleton made up of separate bones, has a spinal column, has a tail, wags the tail sometimes, has a long snout, has a tongue (which sometimes hangs out), can bark, can bite, has a nose with a wet tip, breathes, answers to the name Lucie, identifies my whistle, knows me, loves me, comes when I call her, etc.[13]

11 For example, a personal name (like *Jack*), an identity number (like *ID 4410305013003*), the name of a particular dog (like *Lucie*) or a particular name given to one particular wooden cube of a set of apparently identical looking wooden cubes (like *cube no. 21*).

12 For example, the noun *lion* is the name of a particular type of animal having a particular set of properties which we all know and upon which we all agree; it is not unique enough to be used as a name for a *particular* lion in a circus, since it is a general name.

13 This list can be continued to any degree of completeness, for instance, by listing all the internal organs and their component parts, or by accurately describing all the spots on her skin and their shapes and colours, as well as their exact positions on the dog. It is clearly a big undertaking to set up such a *formal and complete description* of the dog. Fortunately, in real life we have the ability to make a positive identification on information which is formally inadequate (we call it fuzzy information). This is the sort of information which the biologist needs to define his concept of species which we'll meet below. The biologist must find a way to define exactly the minimum properties which he needs to identify a given species absolutely. If he is successful in finding such a minimum set of properties, then the information, even though it is not complete, is not fuzzy.

Embedded in this full set of characteristics are the consecutively smaller subsets of properties which I can easily identify in the full set of characteristics, for example:

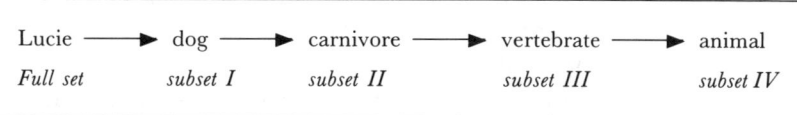

Lucie   dog   carnivore   vertebrate   animal

Full set   *subset I*   *subset II*   *subset III*   *subset IV*

Lucie is a particular dog: she is my dog, and in ordinary life I think that I know all her properties and her complete description; the name Lucie is thus the name of the complete set of properties which I use to identify Lucie when she plays with other dogs in the street; I identify the name Lucie with the object.

If I leave out the property of Lucie that she is my particular dog, I can refer to Lucie as belonging to the subset called dog. If all the particular properties of a dog are left out from the set of properties, then I can refer to Lucie as being a carnivore (Latin *caro/carn-*: flesh; Latin *vorare*: to swallow, to devour), and if I leave that out too, then Lucie is described as just being a vertebrate (having a set of vertebrae or a spinal column). If that is left out too, then all that remains are only those properties which make Lucie an animal (Latin *animal*: an animal, a living creature) and which distinguish her from being a plant, a bacterium, a virus or a fossil.

However, there is nothing in ordinary language and custom that prevents me from inserting as many classes between the concepts Lucie and dog, each with its own set of properties which can be derived from the full set of properties of Lucie, for instance:

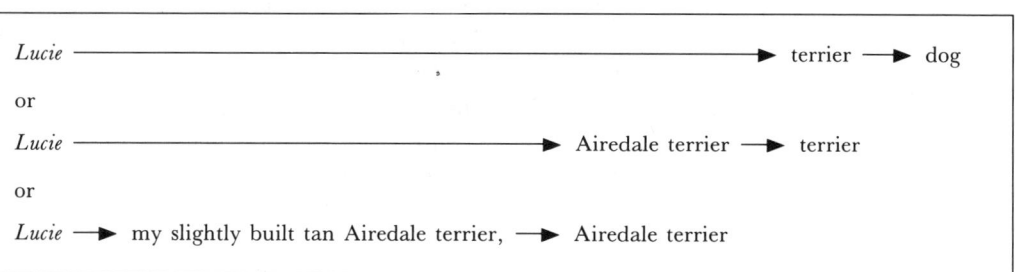

Lucie ⟶ terrier ⟶ dog

or

Lucie ⟶ Airedale terrier ⟶ terrier

or

Lucie ⟶ my slightly built tan Airedale terrier, ⟶ Airedale terrier

and so on. It is clear what is happening: the closer the list of characteristics in the subset approaches the full set of properties, the more elaborate becomes the name of the subset. The next subset name to the right of the full set (for example *Airedale terrier* in this case) is usually extended by adding words before the subset name, such as:

☐ *possessive pronouns* (for example, *my*),

☐ *adverbs* (for example, *slightly*),
☐ *participles* (for example, the past participle *built*), and
☐ *adjectives* (for example, *tan, Airedale*).

This way of speaking about my dog is acceptable and even laudable in ordinary speech, but it leads to great confusion when it is applied to science, since:

☐ it is simply not *precise* enough – it can, therefore, be easily mis-understood (for example, what does 'slightly' mean if one has never seen an Airedale terrier?)
☐ it is not simple
☐ it is too personal and subjective
☐ it cannot be used to generate stable classification schemes which help us to understand nature, because arbitrary subsets can be inserted at any point
☐ it cannot easily be 'internationalised' since it is totally dependent upon the language in use (in this case, English) and the terms used cannot easily be translated into or adapted for other languages.

Clearly, another system is needed which does not suffer from any of the above-mentioned drawbacks and which can be used to name all the animals and plants and other creatures in nature uniquely and in a systematic fashion. Strangely enough, even though the intuitive nomenclature[14] described above is too flexible for scientific use, the method or the set of general assumptions listed above is adequate for developing a scientific system of nomenclature. The point of depar-ture of the method is the scientific description of the object being named in the form of a list of exactly defined properties about which every-body agrees. The more complete these lists of properties are, the bet-ter a scheme of nomenclature or classification can be found. This means, as time proceeds and as science and its methods and appara-tus become more refined, that classification schemes will have to be constantly revised and extended.[15]

Such a system of classification was devised by the Swedish biologist *Linnaeus* in the 18th century, and is known as the *binomial classification system* (Latin *binarius* – *bini,* two by two). This system is mainly based upon the Latin biological vocabulary which will be further elaborated below. It does not try to give names at the so-called personal level which can identify *one* particular living organism, but it attempts to generalise, to categorise, to collect individuals into groups contain-ing similar individuals.

---

14 Latin: *nomen* name; *calare* to call. *Nomenclator,* one who gives a name, one who draws up a classified scheme of names.
15 The history of the classification of living organisms is very interesting; more about this can be found in any good encyclopaedia.

The basic point of departure is the definition of the concept *similar*. Much effort has been expended to give a watertight definition of this concept. The system starts with a collection of similar individuals called a *species* – and the idea is to identify, describe and name every living species. The *species* concept is thus fundamental to the system. Referring to the theory of graphs which we studied at school, we can say that the term *species* refers to the origin of the co-ordinate system in which the particular group of similar individual organisms is going to be classified. One of the main problems with the system is that the origin is not permanent, but as our knowledge of nature increases, and as our instruments with which we probe nature become more and more sophisticated, we collect more and more properties of organisms which may be used to define species. We must, therefore, expect that some species classifications will be revised in the future; as a first step, sub-species of a particular species are defined before a reclassification is attempted.

The classification system makes two main assumptions which are based upon our collective human experience over the ages, namely, that we know that amongst the myriads of organisms on earth there are organisms which we think are so similar to one another that we say that they are indistinguishable from one another on a non-personal level. Such a collection of fundamentally indistinguishable organisms is called a *species* (Latin *species*: appearance, kind, species) and we use the *species*-concept as the starting point of any classification scheme. It is clear that the indistinguishable nature of similar organisms has its origins in the identity of their sets of properties or characteristics which we discussed above. The better their characteristics are listed, the better the starting point of our classification scheme and the more certain we will be that we have really identified the *species* starting-point (which exemplifies the non-personal full set of properties or characteristics) of our classification scheme.

The two assumptions underlying the binomial system are:[16]

☐ every *species* or set of similar organisms can be absolutely identified by means of a name consisting of two parts[17] called the *binomen*:

   – the first part is the generic name or the *genus* name (Latin:

---

16 There are classification problems which cannot be solved by this very simple system of nomenclature, for instance, the variations in plants and trees which are induced by different ecological conditions on the surface of the earth (identical plants may look completely dissimilar). The modern biologist tries to take this into account in his description of plants, and it should ideally be reflected in the nomenclature system.

17 Linnaeus, for instance, replaced the old descriptive name *Genista minima aethiopica foliis thymi confertis* (minute African Genista with the crowded leaves of Thymus) with the systematic **binomen** *Aspalatus thymifolia*.

> *genus/gener-* birth; used as it were, for the surname) which always
> starts with a capital letter;
> – the second part is the *specific epithet,*[18] used as it were, for the
>   Christian name which is usually written with a small letter

☐ every individual organism of a *species* breeds true over considerable periods of time; that is, no short-term evolutionary changes
take place within a particular species[19] so that a stable set of
properties can be found which can be used to define the species
absolutely[20].

It is to be noted that the *species name* consists of the *genus* name followed by the *specific epithet.* The *specific epithet* thus functions as an
inseparable adjective of the *genus* noun in the *binomen* of the species.

In terms of the discussion above, it can thus be said that:

Definition

> the *binomial species name* refers to the full set of non-personal or
> non-individual properties of the plant or animal under consider-
> ation; this set of properties or characteristics is different from
> that of any other species on earth and does not change between
> parent and offspring.

Example

As an example, consider the species name of one of the *Platyhelmintes*
(Greek *platys*[21]: flat; Greek *helmis*: worm), the so-called *flatworms* of
which there are about 15 000 different species (the tapeworms belong
to this group). The species name is *Procotyla fluviatilis,* a freshwater
ciliated (Latin *cilium* plural *cilia:* eyelash – referring to the hairlike
appendages of the flatworm) turbellarian (Latin *turbellae*: a disturbance)
flatworm found in the Great Falls, Virginia, USA. The *genus* name
refers to the Greek *pro*: before, Greek *kotyle*: (drinking) cup, indicating that the worm has a sucking cup at the front which it uses to attach
itself to the host. The specific epithet *fluvialis* refers to the Latin
fluviatilis: of a river; *fluvius* plural *fluvia*: river. This binomen, together

---

18 An *epithet* is an adjective which expresses some real quality or property of the
    object which it describes.
19 This assumption is not quite true, since such random changes do take place in
    some species over shorter periods of time.
20 This assumption is equivalent to the fact that it is assumed that all individuals
    of a particular species are absolutely alike in the properties which are used to
    define the species; they need not look alike in every fashion.
21 Some dictionaries transliterate this Greek word as *platus.*

with the accompanying description or diagnosis (which is not given here), identifies the organism absolutely.

It follows that the *genus* name, when used alone, refers to a subset of the species' properties. Closely-related *species* (plural *species*) are thus collected into a *genus* (plural *genera*). The name of an entity can be linked to other similar biological entities because closely-related *genera* are collected together into a *family* with its own name; closely-related families into an *order*; closely-related orders into a *class*; closely-related classes into a *phylum* (Greek: *phylon* a race); closely-related phyla into a *kingdom*. This constitutes a conceptual classification into graded subdivisions or hierarchy. The complete scientific description of a biological entity is thus the following set of names of which the first two are essential for its absolute identification:

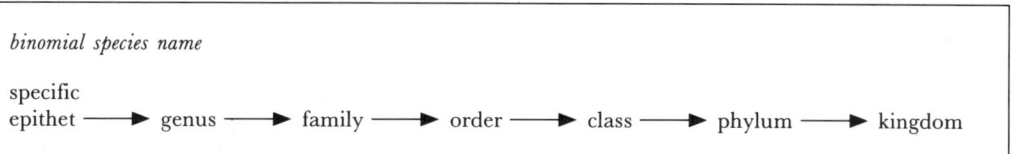

binomial species name

specific
epithet ——▶ genus ——▶ family ——▶ order ——▶ class ——▶ phylum ——▶ kingdom

Going from left to right, we thus see that each following set of properties or characteristics is a subset of the previous (sub)set on its left. This means that the set of properties of a *genus* is common to that of all the species which belong to it[22] and that nowhere on earth does there exist another genus with exactly the same properties: the biological nomenclature system is a mutually exclusive system that gives a unique name to every biological entity. This is shown schematically in the Venn diagram of **Figure 9.3** for a genus with three *species*. From the mathematics which we learnt at school, we remember that this sort of overlapping of sets of characteristic elements is called the *intersection* of the sets I, II and III.

Groups of all sizes (including a variety of a *sub-species*) are called *taxa* (singular *taxon*, from the Greek verb *tassein*, meaning to arrange[23]). The determination of the characteristics of an organism for the purpose of assigning it to a taxon is called *taxonomy* (Latin *nomen*: name). A *taxon* is thus any taxonomic unit, for example, a *species*, a *family*,

---

22 In mathematical language one can say that the set of unique properties of the *genus* is actually the *intersection* of the sets of properties of the species. The same holds for the other subsets, and *mutatis mutandis*, for the other names of the hierarchy listed above.

23 This is also the etymology of our word *taxi* or *taxicab*, meaning 'an arranged vehicle'.

etc. The taxonomical (topological) branching diagram for *genera* consisting of only two species is shown schematically in **Figure 9.4**.

The same type of *topological branching diagram* is found in all schemes in which the properties of something are analysed and catalogued into sets and subsets – it appears in the linguistic analyses of sentences (called *parsing,* from the Latin *pars orationis*: a part of speech), in pure mathematics (the theory of graphs), in the naming of organic compounds, in neural networks, in computer networks, in the study of the symmetry and geometrical shapes of molecules (called *group theory*), and in many more branches of science. Every *node* (Latin: *nodus* a knot) marks a logical branching point where a decision has to be made about the characteristics by means of the question: *does it have the following characteristics, yes or no?* If the answer is *yes,* go this way, and if the answer is *no,* then go the other way.

Figure 9.3
A schematic diagram showing the set of *genus* properties or properties which are common to three species (the *intersection*). Each circle encloses the elements of the set (which are the properties or characteristics of the relevant *species;* in other words, the circle is equivalent to the list of properties). The full *species* name refers to the set of species properties which include the specific epithet and the genus name.

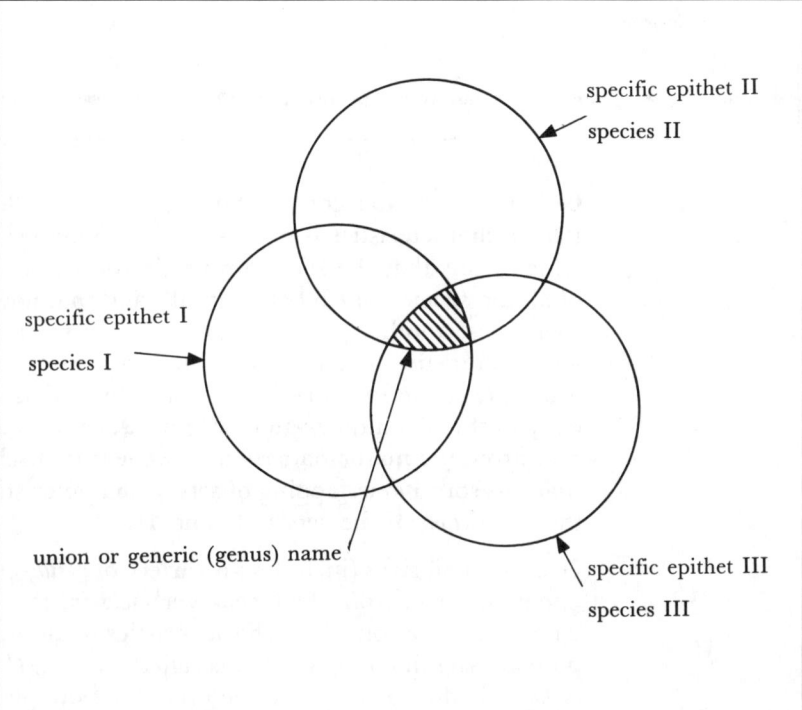

Figure 9.4
The topology of the classification scheme of the biological sciences. In order not to complicate the drawing, it is illustrated for the hypothetical case of only two *species* per *genus*. The topological tree structure is shown only up to the first *class* level; it can, obviously be extended up to the next level of *phyla,* and then on to the *kingdom* level. The names of the species, etc. are all derived from Biological Latin; this will be discussed below.

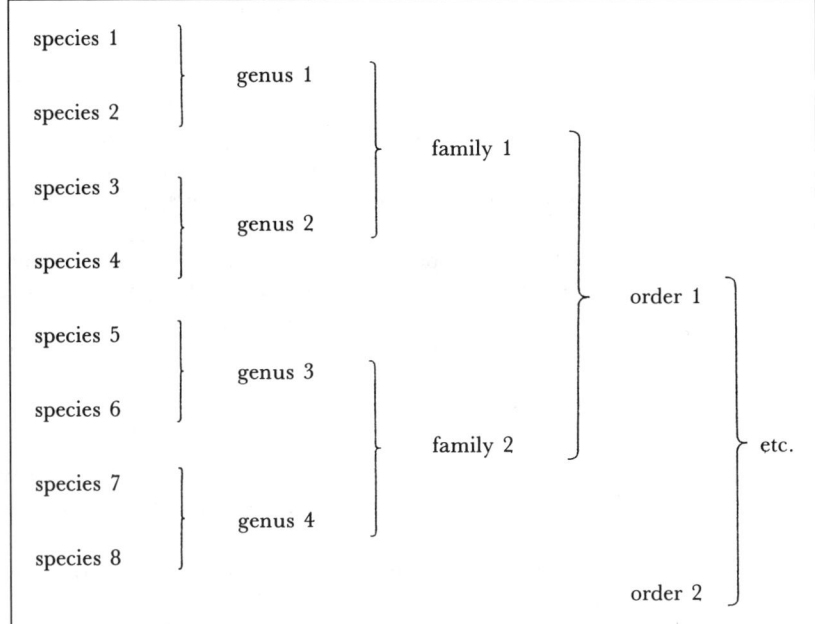

It is not the purpose of this chapter to teach the systematics of plants, animals and other biological entities belonging to the five kingdoms of living nature. The brief introduction above is needed to give the reader a better chance to understand the description of the language which is used to give names to biological entities.

☐ Some Remarks about the Latin of Biology

Latin is still the basis of the language which is needed for the description of plants and animals[24] and every biologist should be familiar with its rudiments, although it is not necessary to study the language itself in all detail – unless one wants to specialise in the classification of biological organisms.

It is usually an aid to memory if one knows what a word means or out of which Latin words it has been put together. It is good practice to try to find out what the *root* of every scientific word means. Once one knows that, then it is much easier to remember the word, to

---

24 This is true not only for the giving of names to biological entities, but also for naming parts of animals, plants and viruses, and even for constructing names to describe experiments and their results.

remember how the thing to which it refers looks, to remember its function, and to remember how to spell it. If one is not very familiar with
Latin, then a good dictionary with etymological entries (Greek: *ety-
mon* true; Greek: *logos* discourse, speech, reason – the true origin
of a word) for every word is indispensable; many good biological books
also give the meaning of words in the text, or in appendices, or in
a glossary (a partial dictionary for specific purposes).[25]

> The most important thing to remember about biological names – for
> organisms, for their organs, for the biochemical and biophysical
> processes which keep them alive, for their reproductive processes, for
> the molecular building blocks which make up their tissues, etc. – is
> that every name has a meaning which can be expressed in everyday
> English.

Latin is a very formal and complex language in which the endings
of nouns, adjectives and verbs indicate their respective functions in
sentences. It is not necessary for the ordinary biologist to learn all
this grammar (it is only needed by the person who wants to specialise in systematics), but one does need a few tips to be able to figure
out the meanings of words, especially if the need arises to look up
the etymology of words in dictionaries. The main emphasis in this
book will be on *nouns* (the names of things); some attention will be
given to *adjectives* and *adverbs* (which qualify verbs and adjectives), but
modern biological Latin verbs will not be treated in detail.

Many examples are cited in this section for the convenience of the
reader. These examples were selected with great care in order to:

☐ give the reader an overview of the most important types of word-
formation in the biological nomenclature which originates in Latin
and Greek

☐ furnish the reader with examples showing how to analyse an
unknown word

☐ give the reader a complete list of the most important Latin and
Greek *prefixes* and *suffixes* used in biological nomenclature, as well
as their ranges of meanings (this list is especially valuable in the
beginning and should be carefully studied)

---

25 A very good book which incorporates all these features is L Margulis and K V
Schwartz, *Five kingdoms – An illustrated guide to the phyla of life on earth*, Second
Edition, W H Freeman and Company, New York, 1988. There are other books,
for instance: D Anderson and R Buxton, *A pocket etymology of medical terms – An
introduction to the Greek and Latin Roots of medical terminology*, Bristol Classical Press,
Bristol University, Bristol, 1981; J A McCulloch, *A medical Greek and Latin work-
book*, Charles C Thomas Publishers, Springfield, Illinois, 1962; G Ahrens, *Natur-
wissenschaftliches und medizinisches Latein*, VEB Verlag Enzyklopaedie, Leipzig,
1977.

☐ provide some of the most important Latin and Greek nouns, verbs, adjectives, etc. which are used in biological nomenclature (this list is not complete – a complete list would fill a large dictionary!).

It is suggested that the reader carefully work through each of these examples in order to develop a feeling for the process of name-giving in biology to animals, plants, organs, processes, etc. This will afford an opportunity to become familiar with the most important Latin and Greek words, their spellings and their meanings. Once these words become familiar, the reader will find that biological nomenclature is no longer something to learn by heart (this includes both the name, as well as the strange spelling of words), but that it becomes a natural way of expressing familiar concepts.

Roots of nouns

Latin nouns are cited in Latin and other dictionaries in a form which does not always show the very important part, called the *root*. A case in point is the Latin word for flesh *caro,* where the root or stem is actually *carn-*. This root appears in the English word which we met above, namely carnivore, flesh-eater. Another example is the Latin word for mouth, *os* (as found, for instance, in the phrase *nil per os* on the notice which is placed in hospitals above patients who are not allowed to eat anything – nothing through the mouth); the root is *or-,* found in our word *oral:*

☐ *oral* adjective, spoken, by word of mouth, not written (*oral examination, tradition*) done or taken by the mouth (*oral dose of medicine*), of the mouth (as used in anatomy).

In this book the Latin noun will be given in the normal dictionary form and the root will be added (if different) after a slash, for example, *os/or-* and *caro/carn-*.

A great number of Greek nouns and some verbs, prepositions and adverbs have found their way into scientific English, and especially into biology. Many of these forms have been latinised and they fit easily into the Latin scheme of things in biology. The Greek alphabet is written differently from ours, but no attempt will be made in this book to write the Greek words in this script; the words will be transliterated into their equivalents in modern English.[26]

If the etymology of a word is not given in an appendix or in a glossary of a textbook of biology, then a normal English dictionary which does not display all the words of every branch of science might be used to derive the meaning of unknown words, provided one knows how scientific words are constructed; some attention will be given to this aspect of the scientific vocabulary below.

---

26 The Greek alphabet is extensively used in mathematics and is displayed in Chapter 10.

In English classes the students were earlier taught the old Edwardian rules of literary English:

☐ prefer the short word instead of the long
☐ prefer the Saxon (or a word derived from old English) word instead of a foreign word, such as those originating from the Romance Languages, or from Greek or Latin.

These 'wise rules' definitely do not apply to science in general and to biology in particular, since biological words are often compounded from other words and are constructed to be both descriptive, as well as to have one clearly-defined meaning; they are, therefore, sometimes very long and complicated-looking. In addition, it is preferred that these words may be understood by all biologists, irrespective of their mother tongue. The ideal of the international bodies which regulate the biological nomenclature is that the nomenclature words be designed in such a way as to be as portable (to use a term relating to computer programs) as possible between the various indigenous languages of the world; this ideal then forces the person who invents a new name to use Latin or Greek as basis. Naturally, these ideals are not easily realised in practice, and especially not in the language which is used by biology, one of the oldest sciences and one with a vocabulary which has developed over a period of more than two thousand years.

Scientific words are sometimes very long, but they are descriptive and precise and they are understood by everybody in the field. Let us look systematically at the construction of some of the words used in biology.

The simplest words

The simplest words are mostly the oldest ones which have been in use for ages and which are taken over into biological English with almost no change from Latin or Greek (or, in rare cases, from some other foreign language). These words are mostly used to describe those parts of the animals or plants which everybody has known since time immemorial, although there are some exceptions. Such words are used in all branches of biology and they are for the most part also found in ordinary literary English, and sometimes also in colloquial speech, for instance, those referring to parts of the human and the animal body. They, therefore, appear in good dictionaries.

A browse through a dictionary of the English language provides many such examples, for instance:

☐ *abdomen* (Latin *abdomen*: belly) *noun,* belly
☐ *antenna* (Latin *antenna*: a yard (of a mast)) *noun,* (in radio and TV) the structure for sending or receiving the electromagnetic signal, (in biology) a feeler or horn of insects, crustaceans or myriopods with which they probe the surroundings

☐ *anterior* (Latin *anterior,* comparative of *ante:* before) *adjective,* before (in time and in place), (botany) towards the edge of the leaf (bract) or away from the axis; *noun,* front side (of the body) – in contrast to the *posterior* (see there)

☐ *cortex* plural *cortices* (Latin: *cortex/cortic-*: the bark of a tree) *noun,* the bark or the skin of a plant; outer part of the brain of mammals

☐ *cranium* plural *crania* (Latin *cranium*: the skull; Greek *kranion*: the skull) *noun,* the skull, the bones enclosing the brain

☐ *dorsal* (Latin *dorsum*: the back) *adjective,* pertaining to the back or back side of some organism where the front and back can be unambiguously identified

☐ *fetus* or *foetus* plural *fetuses* or *foetuses* (Latin *fetus*: offspring) *noun,* the young animal in the egg or in the womb when its parts or limbs have already been distinctly formed by a process of growth

☐ *folium,* plural *folia* (Latin *folium*: a leaf; Greek *phyllon*: a leaf) *noun,* a leaf, a thin plate called a lamina or a lamella

☐ *foramen,* plural *foramina* (Latin *foramen*: an opening or hole made by piercing an object (derived from the Latin verb *forare*: to pierce)) *noun,* a small opening in an object

☐ *forensic* (Latin adjective *forensis*; noun *forum*: marketplace, forum, where the Romans held their courts of law) *adjective,* belonging to the courts of law

☐ *forest* (Latin *forestis*: the wood in the countryside, as opposed to the wood which is planted or cultivated and which was called *parcus* park) *noun,* a large area of natural land covered by trees and underbrush

☐ *forfex* (Latin *forfex/forfic-*: shears, pincers) *noun,* a pair of shears, scissors or pincers

☐ *fork* (Latin *furca*: a two-pronged fork) *noun,* a pronged instrument or anything that divides into branches or prongs; *adjective,*furcate, furcated, forked

☐ *formula,* plural *formulae* (Latin dimunitive of *forma* a shape, a form, a contour, an appearance) *noun,* a prescribed form, a general expression for solving problems (especially in mathematics and physics), a set of symbols in chemistry denoting the elemental composition of a compound, a list of ingredients

☐ *fornicated* (Latin *fornicatus*: arched, or *fornix*: a vault, an arch) *adjective,* arched, used in botany to indicate something which is arching over

☐ *fossa* plural *fossae* (Latin *fossa*: a ditch or furrow) *noun,* a pit or depression, used in anatomy, especially to describe the surface topography of skeletal bones

☐ *fovea* plural *foveae* (Latin *fovea*: a small depression or pit) *noun,* depression or pit

☐ *front* (Latin *frons/front-*: the forehead) *noun,* the forehead, the fore-part of any object

☐ *fungus* plural *fungi* (Latin *fungus*: a mushroom) *noun,* a plant of one of the lowest groups such as mushrooms, toadstools, moulds, etc. which does not contain chlorophyll (Greek *chloros*: pale-green – which is also the origin of the name of the chemical element *chlorine;* Greek *phyllon*: leaf)

☐ *gland* (Latin *glans/gland-*: an acorn) *noun,* a (globular) structure (with or without ducts) in plants and in animals which secretes chemical substances

☐ *graph* (Greek *graphe*: a writing; *graphein*: to write) *noun,* a symbolic diagram, a curve on paper which represents in an analog way the variation of a quantity which is measured or calculated from a formula

☐ *helix* plural *helices* (Greek *helix*: a spiral) *noun,* a screw-shaped coil in three dimensions (in contrast to a *spiral* which is in a plane)[27]

☐ *larva* plural *larvae* (Latin *larva*: a spectre or mask), *noun,* an organism in an active and immature state which eventually goes over into the adult form by a process of metamorphosis, such as occurring in the caterpillar

☐ *lateral* (Latin *latus/lateris*: side) *adjective,* pertaining to the side (of the body)

☐ *mammal* (Latin *mamma*: breast, *mammalis*: of the breast) *noun,* a member of the class *Mammalia* which suckle their young;

☐ *Nemertinea* (Greek *Nemertes*: a sea nymph), *noun,* the ribbon worms which have an anus and a blood vascular system (in contrast to the flatworms)

☐ *nerve* (Latin *nervus*: sinew; Greek *neuron*: a nerve) *noun,* a string-like structure which conducts electrical impulses to and from the brain and from and to the body, respectively

☐ *ocular* (Latin *oculus*: the eye), *adjective,* relating to the eye

☐ *posterior* (Latin *posterior,* comparative of *posterus*: coming after, Latin *post*: after) *adjective,* coming after, later, (botany) on the side next to the axis; *noun,* back side, hinder parts, buttocks – in contrast to the *anterior* (see there)

☐ *solar plexus* (Latin *sol*: sun; Latin *plexus*: plaited, a braid) *adjective* and *noun,* a network of nerves behind the stomach with nerves radiating outward from a central area

---

27 Many dictionaries confuse *helix* and *spiral,* using the word *spiral* also for three-dimensional objects. Many molecules of biological significance, such as DNA, the carrier of genetic material, appear as *helices,* looking like non-rigid cork-screws. It must be remembered that *spirals* and *helices* have a handedness, that is, they can be either right-handed (going around clockwise) or left-handed (going around anticlockwise).

☐ *ventral* (Latin *venter*: the belly) *adjective,* referring to the belly, of the belly, (botany) on the upper side or towards the axis, (zoology) on the side of the animal normally turned towards the ground, and opposite to *dorsal*

☐ *ventricle* (Latin *venter*: the belly) *noun,* any cavity in the body, for instance the cavities or contractile hollow chambers of the heart.

Most of these words appear in modified forms as nouns and as adjectives in English (some even as verbs and adverbs), but all go back to the meanings given here. As an example, consider the noun *graph:*

☐ *graphical* – adjective
☐ *graphically* – adverb
☐ *to graph* – verb.

Many words are derived from the noun *graph,* for instance:

☐ *graphic* – adjective and noun
☐ *graphics* – noun
☐ *graphite* – noun
☐ *grapheme* – noun
☐ *graphemic* – adjective.

These words can also form part of compound words like

☐ *tomography* (Greek *tome*: a cutting) a technique used for producing electronic or computer images representing cross-sections or slices of the body which are indispensable for diagnostic purposes
☐ *photograph* (Greek *phos, photos*: light)
☐ *to photograph*
☐ *bibliography* (Greek *biblion*: a book) used for a list of books (or articles), their authors, and where and when they were published.

The construction of such words is discussed below.

Words compounded from a noun and a verb

Many of the biological words are *compounded* from a Latin or Greek *noun* and a Latin or Greek *verb* to form English nouns, adjectives and verbs.

There are many such words formed from a noun and the Latin verb *ferre* (which has *tuli* as perfect and *latum* as passive perfect; all three appear in words) which means to bear or to carry; the ending *-ferous* is an indication that the word has something to do with *bearing* or *carrying*. Examples of such compound words are, for instance:

☐ *ablate* (Latin *ab*: away from; Latin *ferre/latum*: to carry, to bear) *verb,* to remove or take away from (something)
☐ *circumference* (Latin adverb *circum*: around; Latin *ferre*: to carry, to bear) *noun,* the boundary line, especially of a circle
☐ *fructiferous* (Latin *fructus*: fruit; Latin *ferre*: to bear, to carry) *adjective,* bearing fruit

☐ *furciferous* (Latin *furca*: fork; *ferre*: to bear) *adjective,* bearing a forked appendage or growth

☐ *Loricifera* (Latin *lorica*: corset; Latin *ferre*: to bear, to carry), *noun,* a *Phylum* of the *Animalia,* tiny marine animals with spiny heads and with the abdomen covered with spiny platelets which occur on a girdle or belt, which is called a *lorica*

☐ *seminiferous* (Latin *semen/semin-*: seed; *ferre*: to carry, to bear) *adjective,* seed-bearing (plant)

☐ *septiferous* (Latin *septum* plural *septa*: a partition; Latin *ferre*: to carry, to bear) *adjective,* (something) having partitions

☐ *somniferous* (Latin *somnus*: sleep, *somnium*: a dream; *ferre*: to bring, to carry) *adjective,* (something) causing or bringing sleep

☐ *tentaculiferous* (Latin *tentare*: to feel; Latin *ferre*: to carry, to bear) *adjective,* bearing slender organs for feeling

☐ *transfer* (Latin adverb *trans*: over, across; Latin *ferre*: to carry, to bear) *verb,* to carry or bring over.

Most of these words also appear in scientific English as nouns, adverbs, adjectives and verbs, as the case may be, by appropriately changing their forms; the meanings are, however, always related to those given here.

Other Latin verbs are also common, such as *facere* which has a whole range of meanings around the concepts to make or to produce; it takes the form *-fic* or *-ify* in compound words. Endings *-ify, -fic, -fec(t)* and *-fac,* as well as those starting with *fac-,* are thus an indication that the word has something to do with the action to make something. Examples are:

☐ *chondrify* (Greek *chondros*: a grain, grit, cartilage; Latin *facere*: to make) *verb,* to change into cartilage

☐ *facilitate* (Latin *facile*: easily, readily; *facilis*: easy; *facilitas*: ease, readiness – all derived from Latin: *facere,* to make, to do) *verb,* to make easy, promote

☐ *Foraminafera* (Latin *foramen*: little hole, perforation through a solid object; Latin *ferre*: to bear, to carry) *noun,* a *Phylum* of marine organisms with pore-studded shells of $CaCO_3$, calcium carbonate, which range from the very small to those which are several centimeters in diameter; they attach themselves to the sea sand, or to marine objects

☐ *fructify* (Latin *fructus*: fruit; Latin *facere*; to make) *verb,* to make fruitful; *fructification* (Latin) *noun,* fruit-production

☐ *perfect* (Latin adverb *per:* through; Latin *facere*: to do, to make; *perficere/perfectum* to do well or completely) *adjective,* done thoroughly or completely

☐ *scientific* (Latin *scientia, -sciens -scientis* present participle of *scire*: to know; Latin *facere*: to make) *adjective,* originally, making or producing knowledge, now meaning based on science

☐ *scorify* (Greek *skoria, skor*: dung; Latin *facere*: to make) *verb*, to reduce to *scoria* or dung

Another verb which is used extensively, is the Latin verb *formare* which means: to form. Examples are:

☐ *cruciform* (Latin *crux/cruc-*: cross; Latin *formare*: to form) *adjective*, shaped or formed like a cross

☐ *format* (Latin *formare* to form, from the past participle *formatus*: formed) *noun*, arrangement of data, especially for processing by computer; and *verb*, to arrange data for processing, especially by computer[28]

☐ *fusiform* (Latin *fusus*: spindle; Latin *formare*: to form) *adjective*, shaped like a spindle, that is like a rod, but tapering at both ends.

There are many other Latin verbs which are used in biological terminology (Latin *terminus*: term; Greek *logos*: speech, reason, discourse – science of the proper use of scientific terms, system of terms used in science or in another branch of knowledge) – far too many to list here in all detail. Examples of such verbs (which ought to be learnt since they give access to the meaning of very many other words) and their compounds are, for instance:

☐ *biparous* (Latin *bi/bis*: twice; Latin *parere*: to bring forth, to give birth) *adjective*, bearing two offspring at birth

☐ *bisect* (Latin *bi/bis*; twice; Latin *secare/sectum*: to cut) *verb*, to divide into two (by cutting)

☐ *centripetal* (Greek *kentron*, Latin *centrum*: a sharp point; Latin *petere* to seek) *adjective*, tending towards a centre, towards the middle point of a circular motion

☐ *circumduct* (Latin adverb *circum*: around; Latin *ducere/ductum*: to lead) *verb*, to lead around or about

☐ *Ctenophora* (Greek *kteis*: comb; Greek *pherein*: to bear – the Greek equivalent of the Latin *ferre*) *noun*, one of the kingdom *Animalia*, the so-called sea-jellies, having paddle-like comb-plates, each of which consists of many *cilia* (hairs)

☐ *perforate* (Latin adverb *per*: through; Latin *forare*: to bore, to drill (a hole through)) *verb*, to bore or drill a hole through (something);

☐ *transfuse* (Latin adverb *trans*: across, over; Latin *fundere/fusum:* to pour) *verb*, to pour out (into another container), to transfer (blood to another person's veins).

Again, most of these words can be modified to form verbs, adverbs, adjectives or nouns, as the case may be; they all have meanings derived from those above, and they cover an extensive range of biological terms.

---

28 Not all words starting with *for-* are derived from *formare*.

Connecting vowels -o- and -i-

The compounding of two or more classical Greek and Latin words to form new words usually tries to follow the classical rules of word combinations. There are, however some exceptions. One of these is the insertion of the so-called connecting vowel between two words. With Greek words, this is usually -o-, and with Latin words, it is -i-, although in some Late Latin forms we find that -o- replaces the -i- as a connecting vowel. The connecting vowel makes the word easier to pronounce, even when the first word ends upon a vowel, such as *palaeoichtyology* (which is explained in the first list of this chapter), where two connecting vowels -o- are inserted.

This rule also applies even to words of non-Latin origin, such as *chemotherapy*, where the stem *chem-* comes as an abbreviation of the word *chemistry* (Arabic origin), and *therapy* from the Greek *therapeia*: service, treatment. Another Latinised example is *chemisorption*, where the Latin connecting vowel -i- is inserted between the stem *chem-* and the Latin *sorbere/sorbtum*: to suck in. The stem *chem-* gives rise to both *chemo-*, as well as *chemi-*, but there are no general rules to tell which is used when. Examples are:

- [ ] *chemotropic* (Greek *tropos*: a turning) *adjective,* response of an organism caused by a chemical irritant which may manifest itself in differential, biased or stunted growth
- [ ] *chemotaxis* (Greek *tassein*: to arrange) *noun,* the response of a whole organism to a chemical stimulant.

Just to make the picture complete, the connecting vowels are usually not inserted before a vowel when the word is easy to pronounce:

- [ ] *chemurgy* (Latin *urgere*: to press; Greek *-ourgia*: working) *noun,* a branch of chemistry which deals with the use of agricultural and other raw materials for the chemical industry.

Words compounded from a noun and another noun

Many English biological nouns and adjectives are compounded from two Latin nouns: or from two Greek nouns or from a mixture of Greek and Latin (the international scientific bodies disapprove of the practice of mixing Greek and Latin and actively discourage it, but many of the older terms formed in this way are still in use and will be used for a long time):

- [ ] *actinopodia* (Greek *actinos*: ray; connective vowel: -o-; Greek *pous*: foot) *noun,* a *Phylum* from the kingdom *Protocysta* (Greek *protos*: very first; Greek *khystos*: to establish) which includes the algae and the water moulds − the organisms have structures radiating outwards, and a unique radially-symmetrical skeleton of $SrSO_4$ or strontium sulphate crystals (see **Figure 9.11**)
- [ ] *blastopore* (Greek *blastos*: a sprout; Greek *poros*: a passage) *noun,* the orifice of an object

☐ *Dermaptera* plural *Dermapterae* (Greek *derma*: skin; Greek *a-*: privative word, indicating absence or privation; Greek *pteron*: wing; English adjective *apterous* means without wings); *noun,* an order of insects with forewings (when present) in the form of firm *elytra* (Greek *elytron*: a sheath), that is, the fore-wing of the insect (beetle) is modified to form a skinlike sheath for the hind-wing – the word thus means something like skin, but without being a wing

☐ *Euglenophyta* (Greek *eu*: true; Greek *glene*: eyeball, socket of a joint; Greek *phyton*: plant), *noun,* a *Phylum* of the kingdom *Protoschista,* which is in-between being a plant or an animal, for it contains chlorophyll in *chloroplasts* which photosynthesise, but it can sense light with an eyespot and can move with an anteriorly attached *undulipodium* (Latin *unda*: a wave; Latin *undulatus*: undulate; Greek *pous/podos*: a foot)

☐ *fumitory* (Latin *fumus*: smoke; Latin *terra*: earth, soil) *noun,* a plant from the genus *Fumaria* which is akin to the poppies (the connection between the name of the plant and the etymology of the word is not clear – maybe they are the first plants to come up after the fields have been burnt?)

☐ *helianthemum* (Greek *helios*: sun; Greek *anthos*: flower) *noun,* the rockrose genus; *heliantus* (same etymology) *noun,* the sunflower genus

☐ *hemisphere* (Greeke *hemi*: half; Greek *sphaira*: sphere) *noun,* a half sphere[29]

☐ *Hemiptera* (Greek *hemi*: half; Greek *pteron*: a wing) *noun,* a class of insects with wings

☐ *myograph* (Greek *mys, myos*: muscle; Greek *graphe*: a writing, *graphein* to write[30]) an instrument for recording muscular contractions.

The prefixes: words mostly compounded from prepositions or adverbs and nouns or verbs

There are many *prepositions* (which govern *nouns*) and *adverbs* (which govern or modify verbs and adjectives) in Latin and Greek which have found (and are still finding) their way into the compound nouns, adjectives and verbs of biological terminology. Such words are very valuable to know by heart, since they appear everywhere, but care has to be taken, because some of them have a range of meanings (like *dia-, kata-, meta-* and *para-*) and some may even be confused (such as the prefixes *ana-* and the compound prefix *an + a-*); these cases are discussed below. These adverbs and prepositions mostly append to other words in the form of prefixes (which is formed from the Latin adverb *pre* meaning before and the Latin verb *figo* – *figere* – *fixtum,* meaning to fix, to fasten, to attach).

---

29 There are very many words incorporating the Greek word *hemi-* half, such as *hemicrania,* a headache confined to one side of the head or cranium; *hemicrystalline,* consisting of crystals in a (partly) glassy groundmass, etc.

30 There are very many words of which *graph* forms a part – some of them were mentioned above.

A preposition is a word placed before a noun or its equivalent in order to indicate some relation. An adverb functions in the same way with respect to a verb or to an adjective. These Latin and Greek prepositions and adverbs act as *prefixes* to other words and are very popular in literary and in colloquial English – just think of all the hundreds of words which start with the prefixes *ad-, per-, pre-, pro-, super-, inter-* and *intra-*, for instance. The same set of prefixes is also used in biological terminology and the most important of them are collected in the list below which can be used as an aid to dissect an unknown word into its parts in order to make it comprehensible. Most of the prefixes can function either as prepositions (to nouns) or as adverbs (to verbs and to adjectives); their functions are, therefore, not further indicated in the list.

a- or ab- (before a vowel): (Latin) from, away from, of;

☐ *aberrant* (Latin *ab-*: away from; Latin: *errare* to wander, to stray, to lose one's way, to make a mistake) *adjective,* straying from a (moral) standard, deviating from the usual type.

a- or an-: Greek privative word indicating privation or deficiency of something;

☐ *achlorhydria* (Greek *a-*: privative word; Greek *chloros*: pale-green; Greek *hudor*: water) *noun,* absence of hydrochloric acid (in the gastric juices)

☐ *achromatic* (Greek *a-*: privative word; Greek *chroma*: color) *adjective,* (biology) staining without color in histology, colorless; (optics) without any refractive colours, that is, applied to a lens in which the chromatic aberration has been corrected

☐ *adiabatic* (Greek *a-*: privative word; Greek *dia*: through; Greek *bathos*: passable) *adjective,* (thermodynamics) without passing (heat) through (a barrier)

☐ *adiathermic* (Greek *a-*: privative word; Greek *dia*: through; Greek *therme*: heat) *adjective,* impervious to heat; without passing heat through (a barrier)

☐ *asymmetrical* (Greek *a-*: privative word; Greek *syn*: together; Greek *metron*: a measure) *adjective,* not symmetrical, without any symmetry, in biology often referring to the absence of the plane of symmetry that goes with the phenomenon, plural phenomena, (Greek *phainomenon*, present participle of the verb *phainein*: to show) of bilateralism or two-sidedness or right- and left-handedness

☐ *anaemia* (Greek *an-*: privative; Greek *haima*: blood) deficiency of blood, or of red blood corpuscles.

ad- = ac- (before c) = af- (before f) = ag- (before g): = al- (before l) = an- (sometomes before n) = ap- (before p) = ar- (sometimes before r) = as- (before s) = at- (before t): (Latin ad) to, towards, near

☐ *adrenal* (Latin *ad*: near; Latin *renes*: kidneys) *adjective*, near the kidneys; *adrenalin* (the suffix *-in* commonly indicating in science that a compound is secreted by a gland) *noun*, hormone secreted by the glands near the kidneys, the adrenal glands

☐ *adnate* (Latin *ad*: to; Latin *natus*: born) *adjective*, (botany) attached to another organ (especially by the whole length)

☐ *allatus* (Latin *ad* = *al*: towards; Latin *ferre*: to bring, to carry, perfect participle *latus*: brought) *adjective*, brought towards

☐ *appendix* (Latin *ad* = *ap*: to, near; Latin *pendere*: to hang); *noun*, something that hangs on or near (something else)

☐ *assurgens* (Latin *ad* = *as*: to, towards; Latin *surgere*: to rise, present participle *surgens*: rising) *participle*, rising towards

☐ *attingens* (Latin *ad* = *at*: towards; Latin *attingo*: to reach, present participle *attingens* reaching; Latin *tangere*: to reach, to touch) *participle*, reaching towards.

ambi-: (Latin, found only in compound words) away from, used also in the sense of 'equally well' and 'on both sides';

☐ *ambivalent* (Latin *ambi-*: on both sides, equally well; Latin *valens/valent-* present participle of *valere*: to be strong) *adjective*, (psychology) having opposing feelings about something.

ana-: (Greek) up, again, according to, throughout, back, backwards;

☐ *Anabas* (Greek *ana*: up; Greek *banein* to go), *noun*, a climbing genus of fish, such as the climbing perch that often leaves the water

☐ *Anableps* (Greek *ana:* up; Greek *blepein*: to look) *noun*, a genus of fish having open air bladders and projecting eyes which are divided into two parts for simultaneous vision in air and water, that is, it can look up, while looking down

☐ *anabolism* (Greek *ana*: up; Greek *bole*: a throw) *noun*, synthesis of chemical substances by the cell protoplasm

☐ *Anacardium* (Greek *ana*: according to; Greek *kardia*: heart) *noun*, the genus of the cashew-nut, the fruit having the shape of a heart; *Anacardiaceae* (the ending *-ceae* indicating a name).

ante-: (Latin) before; (do not confuse with *anti*);

☐ *antenatal* (Latin *ante*: before; Latin *natus*: born) *adjective*, before birth.

anti-: (Greek) against, instead of (do not confuse with *ante*); see also *contra-*;

☐ *anthelmintic* (Greek *anti:* against; Greek *helmins, helmintos*: a worm) *adjective*, destroying or expelling worms; *noun*, the medicine which destroys (intestinal) worms

☐ *antibiotic* (Greek *anti*: against; Greek *bios*: life) *adjective*, inhibiting life; *noun*, the medicine which inhibits life (of certain bacteria only, ideally without affecting the host)

☐ *antibody* (Greek *anti*: against; English: body) *noun,* a defensive pro-
tein produced in the body in response to an attack by a foreign
body, such as the toxin of a bacterium or virus or other parasite

☐ *antigen* (Greek *anti*: against; Greek *gennaien*: to engender, to beget,
to sow the seeds of, to produce) *noun,* any substance or toxin from
an organism which stimulates the production of the *antibody* in the
host.

**cata-: (Greek *kata*) down, thoroughly or completely, against, back,
(arrange) in order;**

☐ *catalyst* (Greek *kata*: completely; Greek *lyein*: to loosen) *noun,* a chemical substance accelerating a chemical reaction without suffering
any chemical change itself at the end of the reaction

☐ *cataphyll* (Greek *kata*: down; Greek *phyllon*: a leaf) *adjective,* (botany)
a simplified or rudimentary leaf (contrasting with *euphyllum*: a true
leaf)

☐ *catarrh* (Greek *kata*: down; Greek *rheein*: to flow) *noun,* a discharge
of a watery fluid by the mucous membranes of the nasal passages,
commonly caused by a cold.

**con- = com- = co- = cor- = cum-: (Latin *cum*, becomes *com-*
before a labial consonant, such as *b, p, m*; changes into *co-*
before a vowel or *h*; changes into *cor-* before *r*) with, together;
this prefix corresponds to the Greek *syn-* (see there);**

☐ *coalitus* (Latin *cum* = *co-*: together; Latin *alere* to feed, to nourish,
to support, to sustain, to maintain and to strengthen, the perfect
participle being *alitum/altum*) *adjective,* united by growth, sustained
together

☐ *commix* (Latin *cum*: with; English: *mix* which is derived from the
Latin *miscere/mixtus*: to mix) *verb,* to mix together

☐ *compute* (Latin *cum*: with; Latin *putare*: to reckon) *verb,* to calculate
with numbers

☐ *concatenate* (Latin *cum* with; Latin *catena*: chain) *verb,* to link together
in a chain

☐ *corrasus* (Latin *cum* = *cor-*: together; Latin *radere*: to scrape, to shave,
to rub or to smooth, participle *rasum*: shaved[31]) *adjective,* scraped
together.

contra-: (Latin) against, opposite, the other side; see also *anti-*;

☐ *contra-indicate* (Latin *contra*: against; Latin *indicare*: to make known)
verb, to make known against, to say that some (medicine) is unsuita-
ble (in some cases because of an allergic reaction by the patient)

☐ *contraceptive* (Latin *contra*: against; Latin *concipere/conceptum*: to conceive) *noun,* (chemical) substance or object preventing pregnancy.

---

31 This is the origin of our word *razor,* an apparatus which scrapes and smooths
the skin of the face.

de-: (Latin) from, away from, down from;

☐ *deglutination* (Latin *de*: down; Latin *glutare*: to swallow) *noun*, the act of swallowing down (something)

☐ *dement* (Latin *de*: from; Latin *mens/ment-*: mind; Latin *demens/dementis* out of one's mind) *transitive verb*, to drive crazy or insane.

dia-: (Greek) through, apart, asunder;[32]

☐ *diaphragm* (Greek *dia*: through; Greek *phragma*: a fence) *noun*, a thin separating membrane, the midriff (which separates the abdomen from the chest)

☐ *diarrhoea, diarrhea* (Greek *dia*: through; Greek *rhoia*: a flow) *noun*, persistent loosening of the bowels; watery *faeces*

☐ *diastole* (Greek *dia*: asunder; Greek *stellein*: to place) *noun*, the dilation of the heart

☐ *diatom* (Greek *dia*: through; Greek *temnein*: to cut; Greek *diatomos*: cut through) *noun*, a class of microscopic unicellar algae with silicified shells consisting of two halves.

dys-: (Greek) found only in compound words which have a connection with the concepts difficulty, be bad, be unfavourable;

☐ *dysentry* (Greek *dys*: ill, bad, amiss; Greek *enteron*: intestine) *noun*, a sickness caused by the infection with *Entamoeba histolytica*, causing inflammation of the colon and diarrhoea

☐ *dysfunction* (Greek *dis*: ill, bad; English: *function*) *noun*, the impairment of the functioning of any organ or thing

☐ *dyspepsia* (Greek *dys*: ill, bad; Greek *pepsis*: digestion) *noun*, indigestion.

e- or ex-: (Latin) out of, from;

☐ *emit* (Latin *e*: out of; Latin *mittere*: to send) *verb*, send out (light, sound)

☐ *experiment* (Latin *ex*: from; old Latin verb *periri*: to try) *noun*, something done to test a theory; *verb*, to do something to test a theory or conjecture or to discover (something)

☐ *explode* (Latin *ex*: from; Latin *plaudere*: to clap the hands) *verb*, to burst with a big bang.

ecto-: (Greek *ektos*) outside;

☐ *Ectoprocta* (Greek *ectos*: outside; Greek *proctos*: anus), *noun*, a *Phylum* of the Animalia which look very similar to the seaweeds or moss, with C-shaped bodies, the mouth and the anus facing up to the surface of the water.

---

32 Care must be taken not to confuse the Greek prefix *dia* with the word *di:* two, twice, which is followed by a word which starts with an *a-*, for instance, *diacid:* an acid with two replaceable hydrogen atoms; *diatomic*, consisting of two atoms.

endo-: (Greek) inside, within (contrasting with *exo-* or *ecto-*);

☐ *Aeroendospora* (Greek and Latin *aer*: air; connective *-o;* Greek *endos*: within, inside; Greek *spora*: a seed) *noun,* a *Phylum* of aerobic endospore-forming bacteria, that is, the bacteria live in an oxygen-rich atmosphere and form spores inside themselves.[33]

exo-: (Greek) outside (contrasting with *endo-*, and similar to *ecto-*);

☐ *exodermis* (Greek *exo*: outside; Greek *dermis*: skin) *noun,* the outer cortex layer of a root, the outside skin

☐ *exogen* (Greek *exo*: outside; Greek *gennein*: to produce) *noun,* a dicotyledon which is called by this name because its stem thickens by growing layers on the outside of the inner wooden part.

extra-: (Latin) outside;

☐ *extracranial* (Latin *extra*: outside; Latin *cranium*: skull) *adjective,* outside the skull

☐ *extrorse* (contracted from the Latin *extra*: outside; Latin *tortus*: turned) *adjective,* (botany) turned outward as in an anther (that part of the stamen which contains the pollen) which opens outward towards the outside of a flower.

hama-: (Greek) together with, together (rare);

☐ *Hamamelis* (Greek *hama*: together with; Greek *melon*: apple) *noun,* an American witch-hazel genus: a small shrub similar to the apple (called a medlar)

☐ *hamarthritis* (Greek *hama*: together; Greek *arthron*: a joint; Greek *arthritis*: gout, inflammation of a joint) *noun,* gout or inflammation in all the joints of the body (= *panarthritis*)

hyper-: (Greek) above, over, too much or too great;

☐ *hyperpyrexia* (Greek *hyper-*: too great; Greek: *pyretos* fever; Greek *pyr*: fire; *pyrexia*: fever) *noun,* an abnormally high fever or very high body temperature.

hypo-: (Greek) below, beneath; deficient, defective, inadequate;

☐ *hypoblast* (Greek *hypo*: beneath; Greek *blastos*: bud) *noun,* the inner germ-layer of a gastrula (derived from the Greek *gaster*: belly) *noun,* the name of an embryo at the stage where it forms a two-layered cup by the folding-in of its wall

☐ *hypodermic* (Greek *hypo*: beneath; Greek *derma*: skin) *adjective,* referring to something that is or takes place under the skin (equivalent

---

33 This word has a **connective vowel** *-o-* inserted before the word *endo-* simply because it is easier to pronounce with the extra *-o-;* this seems to contradict the rules for the insertion for the connective *-o-* as given above. The reason for this is found in the fact that the initial *e-* of a Greek word is pronounced as *he-* (as, for instance, found in the English transliteration *Helena,* for *Elena,* the classical Greek female name).

to: *subcutaneous*, Latin *sub*: under; Latin *cutis*: skin); hypodermic is also used colloquially as a *noun* to indicate a hypodermic needle.

in- or im-: (Latin) used actively to indicate motion *into*, or passively, to indicate that something is inside; *in-* = *un-* in some cases, meaning 'not';

☐ *immerse* (Latin *in*: into; Latin *mergere*: to plunge) *verb*, to place or dip under the surface of a liquid, to plunge into a liquid, to apply a liquid to the object-glass of a microscope to reduce the refraction of the specimen

☐ *incontinence* (Latin *in*: not; Latin, *continere*: to keep together, to preserve, to retain – the present participle is *continens*: retaining, preserving) *noun*, the inability to retain something (such as *faeces* or urine).

inter-: (Latin) between;

☐ *intercostal* (Latin *inter*: between; Latin *costa* plural *costae*: rib) *adjective*, (the tissues and muscles) between the ribs

☐ *interramal* (Latin *inter*: between; Latin *ramus* plural *rami*: a branch) *adjective*, situated between the branches of something

☐ *intersect* (Latin *inter* between; Latin *secare*: to cut – perfect passive participle *sectum*: cut) *verb*, to cut across one another (for instance lines), to divide something into parts.

intra-: (Latin) within, inside;

☐ *intra-abdominal* (Latin *intra*: inside, within; Latin *abdomen/abdomin-*: belly) *adjective*, inside the abdomen, belly.

meta-: (Greek) among, with, beside, after;

☐ *metamorphosis* plural *metamorphoses* (Greek *meta*: after; Greek *morphe*: form) *noun*, the phenomenon when an organism changes into another form (such as worm: pupa: moth)

☐ *meta-* as prefix to the name of a benzene-like compound which refers to a substituent in the third position in a ring (in the sequence: first position, second position called *ortho* or *o-*, third position called *meta* or *m-* (beside, next to), and the fourth position called *para* or *p-* (by the side, beside).

ortho-: (Greek) straight, upright, right;

☐ *ortho-*: the substituent adjacent to the first in a benzene ring; see *meta-*

☐ *orthognathous* (Greek *ortho*: right, straight; Greek *gnathos*: jaw) *adjective*, having a jaw that is normal, that is which neither recedes nor protrudes

☐ *Orthoptera* (Greek *ortho*: straight; Greek *pteron*: wing) *noun*, name of an order of insects (for instance, cockroaches) in which the firm fore-wings serve to cover the folded hind wings.

para-: (Greek) beside, by the side; faulty, abnormal;

- [ ] *para-*: see *meta-* above for prefix to benzene-like compounds;
- [ ] *paracusis* (Greek *para*: beside; Greek *acousis*: hearing) *noun,* disordered hearing
- [ ] *paraheliotropic* (Greek *para*: beside; Greek *helios*: sun; Greek *trepein*: to turn) *adjective,* turning towards the sun
- [ ] *paraligula* (Greek *para*: beside; Latin: *ligula* dimunitive of *lingua*: tongue) *noun,* any one of the two appendages at the ligula (tongue; lower labium or lip) of some insects.

per-: (pel- before *l*, very rare): (Latin) through; also indicating an excess (in modern science)

- [ ] *pellucid* (Latin *per*: through; Latin *lucere*: to shine) *adjective,* clear, transparent to light
- [ ] *periodate* (Latin *per*: intensive word; Greek *ioeides*: violet-coloured; chemical term *iodate*: containing the IO_3-ion) *noun,* containing an excess of oxygen, that is the IO_4-ion
- [ ] *permeate* (Latin *per*: through; Latin *meare*: to pass) *verb,* to pass through (the pores of something), to penetrate and fill (the pores of something).

peri-: (Greek) around;

- [ ] *Periophthalmus* (Greek *peri*: around: Greek *ophthalmos*: eye) a genus of fishes with protruding mobile eyes
- [ ] *perisperm* (Greek *peri*: around; Greek *sperma*: seed) *noun,* (botany) nutritive tissue in a seed around the nucleus
- [ ] *periosteum* (Greek *peri*: around; Greek *osteon*: a bone) *noun,* the protective tough fibrous membrane covering a bone.

post-: (Latin) behind, after;

- [ ] *post-ocular* (Latin *post*: behind; Latin *oculus*: the eye) *adjective,* situated behind the eye
- [ ] *post-mortem* (Latin *post*: after; Latin *mors/mort-*: death, used in the accusative *mortem*) *adjective* and *noun,* after death.

pre- or prae-: (Latin *prae*) in front of;

- [ ] *praenomen* (Latin *prae*: in front of; Latin *nomen*: name) *noun,* the generic name of a species
- [ ] *prepollex* (Latin *prae*: in front of; Latin *pollex/pollic-*: thumb) *noun,* a rudimentary innermost finger.

pro-: (Latin, Greek) before (in time and in place), beforehand, (in new words) in favour of;

- [ ] *procerebrum* (Latin *pro*: before; Latin *cerebrum*: brain) *noun,* forebrain
- [ ] *proglottis* plural *proglottides* (Greek *pro*: before; Greek *glottis* plural

glottidos: a pipe or flute mouthpiece which can be detached) a detachable tapeworm segment.[34]

re-: (Latin) before, in front of; also indicates intensifying an action or repeating it;[35]

☐ *rebloom, reblossom* (Latin *re*: intensifying, repeating; Latin *flos/flor*-: flower, and old English *blostm/blostma*) *verb,* to flower again.

retro-: (Latin adverb) backwards, on the back side, behind;

☐ *retrobulbar* (Latin *retro*: behind; Latin *bulbus*: onion) *adjective,* behind the eyeball
☐ *retromingent* (Latin *retro*: backwards; Latin *mingere*: to urinate) *adjective,* applied to an animal which urinates backwards.

semper-: (Latin adverb) always;

☐ *sempervirens* (Latin *semper*: always; Latin *virere*: to be green, to be blooming, present participle *virens*: greening, blooming) *adjective and noun,* evergreen.

sub-: (Latin) under;

☐ *subcortical* (Latin *sub*: under; Latin *cortex/cortic-*: bark (of a tree)) *adjective,* under the cortex or bark.

super-: (Latin; see also Greek *hyper*) used actively, to indicate motion upwards, or passively, to indicate that something is already up, above or over something else; over, above, from above, during, concerning, beyond, more than;

☐ *supersaturation* (Latin *super*: over and above, more than; English *saturation*, derived from Latin *saturare*: to fill) *noun,* a phenomenon in which more than the quantity of solute which is normally expected to dissolve in a solvent at a particular temperature has actually dissolved
☐ *superfetation* (Latin *super*: more than; Latin *superfetare*: to conceive again while still pregnant; Latin *fetus*: fetus) *noun,* the phenomenon of conceiving again while still pregnant.

---

34 This word is not derived from the English word *glottis,* which, in turn is derived from the Greek word *glottis* plural *glotta*: the tongue.

35 The intensifying and repeating *re-* can be added to very many words, and the meaning of such words can always be found by looking up the second part of the compound word in a dictionary. For instance, the noun *recrystallisation* is easily found by looking up *crystallisation,* and it means a *process* in which something is crystallised a second time. This intensifying or repeated aspect of the prefix *re-* must never be forgotten, since it implies that something has already been done, and that it must be repeated. This is the case with respect to the instruction to *recrystallise* something. This means that the first crystallisation from the solution did not yield the pure compound in the form of pure crystals, so that the crystals must be dissolved again in the solvent and the crystallisation be repeated at least once.

supra-: (Latin) above;

☐ *supracostal* (Latin: *super* above; Latin *costa* plural *costae*: rib) *adjective,* above a rib.

syn- or sym-: (Greek) together, together with, with; this prefix corresponds with the Latin prefix *con-* = *cum-* = *co-* = *cor-* = *com-* (see there);

☐ *synapse* (Greek *syn*: together; Greek *haptein*: to fasten) *noun,* the place where the pairing, joining or interlacing of nerve-cells takes place

☐ *synapsis* (Greek *syn*: together; Greek *haptein*: to fasten) *noun,* the pairing of chromosomes from the paternal and the maternal sides before the division takes place

☐ *syncarpous* (Greek *syn*: together; Greek *karpos*: a fruit) *adjective,* (botany) having united carpels or fruit

☐ *synchondrosis* (Greek *syn*: with; Greek *chondros*: a grain, a grit, a cartilage) *noun,* connection of bones (vertebrae) by cartilage

☐ *synergy* (Greek *syn*: together, Greek *ergon*: work) *noun,* combined or co-ordinated action or process.

trans-: (Latin) across, beyond;

☐ *transient* (Latin *trans*: across; Latin *ire*: to go, present participle *iens*: going) *adjective,* passing, of short duration; *noun,* something which passes very quickly across (for example a signal flash across the screen of an oscilloscope)

☐ *transmute* (Latin *trans*: across; Latin *mutare*: to change) *verb,* to change to another form, shape or element

☐ *transverse* (Latin *trans*: across; Latin *vertare*: to turn) *adjective,* set cross-wise, crossed, across.

The prefixes: the numerical words

There are very many Greek and Latin counting adjectives and adverbs added in front of words to indicate number. This is done in four ways and they can be distinguished by four questions:

☐ *How many?* – the numeral used is the *cardinal number,* such as one, two, three, etc.; in this case fractions of numbers can also be used, such as half, quarter, etc.

☐ *Which in order of number?* – the numeral used is the *ordinal number,* such as first, second, third, etc.

☐ *How many each?* - the numeral used is the *distributive number,* such as one each, two each, etc.

☐ *How many times?* – the numeral used is the *numerical adverb,* such as once, twice, thrice, five times, etc.

The most important of these counting-word prefixes are listed below in alphabetical order, together with their meanings. Some of the usage in modern biological nomenclature is somewhat different from the dictionary Latin forms, since these words are sometimes not used in their correct linguistic forms.

bi-: (Latin, derived from adverb *bis*) double;

☐ *biped* (Latin *bi-*: two; Latin *pes/pedis*: foot) *noun,* a two-footed animal.

bini- or bin- (before a vowel): (Latin) two each, two-by-two; (very rare)

☐ *binaural* (Latin *bi-*: two; Latin *auris*: ear) *adjective,* having two ears.

bis-: (Latin) twice;

☐ *biscoctiformis* (Latin *bis*: twice; Latin *cocquo/coctus*: to cook; Latin *forma*: shape; *formare*: to shape; Medieval Latin *biscoctus*: biscuit – this word is the origin of our word biscuit, or something that is baked twice) *adjective,* shaped like a (finger) biscuit.

centi-: (Latin) one hundred;

☐ *centipede* (Latin *centum*: one hundred; Latin *pes/pedis*: foot) *noun,* any one of the myriapods (Greek *myrios*: numberless; Greek *pous/podos*: a foot).

deci-: (Latin) tenth;

☐ *decimeter* (Latin: *deci-*: tenth; Greek *metron*: a measure) *noun,* one tenth of a meter.

demi-: (French, derived from Latin origin: *medius*: the middle) half;

di-: (Greek, related to *dis* twice) double, twice;

☐ *dicotyledon* (Greek *di*: twice; Greek *coteledon*: a cup), *noun,* a plant of the *Angiosperms,* having its embryo in two cups or in two halves.

dicha- = dicho-: (Greek) in two;

duo-: (Latin) two;

hemi-: (Greek) half (found only in compound words);

hexa-: (Greek) six, sixfold;

milli-: (Latin, derived from *millesimus*: a thousandth) one thousandth, or .001 of something;

mono-: (Greek adjective *monos*: single) single, one only;

☐ *monoblepsis* (Greek *monos*: single; Greek: *blepsis*: sight) *noun,* a condition in which the one eye sees better than both together.

multi-: (Latin) many;

☐ *multiparous* (Latin *multi*: many; Latin *parere*: to give birth) *adjective,* used for an animal which gives birth to many young animals at the same time (such as a rabbit).

oct-, octa-, octo-: (Latin and Greek) eight, eightfold;

☐ *octopod* (Latin *octo*: eight; Greek *pous, podos*: foot) *noun*, eight-armed *cephalopod* (Greek *kephale*: head; Greek *pos, podos*: foot; connective *-o-*: an animal having a foot on the head, that is, tentacles or 'arms' directly attached to the head) belonging to the order *Octopoda* (for instance, an octopus).

pent-, penta-: (Greek) five, fivefold;

☐ *pentagon* (Greek *pente*: five; Greek *gonia*: angle) a rectilineal plane figure having five corners (angles) and five sides

☐ *Pentstemon* (Greek *pente*: five; Greek *stemon*: warp or to become perverted or distorted – the Greek word *stemon* is reminiscent of the Latin word stamen plural *stamina*: a warp or upright thread of a loom, used for the stamen of a flower), *noun*, a North-American genus of flowering plant in which the fifth *stamen* is sterile.

semi-: (Latin) half;

☐ *semidiurnal* (Latin *semi-*: half; Latin *dies*: day; *diurnus*: of or belonging to the day, that is, a 24 hour period) *adjective,* accomplished in half a day or in 12 hours.

septem-, sept-, septen-, septi-: (Latin) seven, sevenfold, seventh;

☐ *septilateral* (Latin *septem*: seven; *latus/later-*: side) *adjective,* seven-sided.

sex-, sexa-, sexi-: (Latin) six, sixth;[36]

☐ *sexlocular* (Latin *sex*: six; Latin *loculus,* the dimunitive of *locus*: a place) *adjective,* having six compartments.

ter-: (Latin) thrice, three times;

☐ *tergerminate* (Latin *ter*: three times; Latin *germinare/germinatus*: to sprout) *adjective,* as when a petiole (leaf-sprout[37]) bears at its top end two little leaves, between which arise two secondary petioles, each bearing two leaflets at its tip.

tri-: (Latin, Greek) three;

☐ *triaxial* (Latin *tri-*: three; Latin *axis*: axis) *adjective,* having three axes

☐ *tristichous* (Greek *tri-*: three; *stichos*: a row; Greek *tristichia*: a triple row) *adjective,* in or having three rows.

uni-: (Latin) consisting of only one;

☐ *unilabiate* (Latin *uni-*: one; Latin *labium*: lip) *adjective,* one-lipped.

36　The English word *sex* is derived from the Latin noun *sexus*: sex of male and female (in the beasts).

37　The same word *petiole* (derived from the Latin *petiolus*: a little foot) is also used in zoology, but it indicates a stalk-like structure, such as the abdomen of a wasp.

The suffixes: words added at the end of other words to extend their meanings

There are many suffixes (Latin *sub-* = *suf-*: under) which are added at the ends of other words to extend or change their meanings. Some of these endings are derived from Greek or Latin words, but others are recent creations which bear no resemblance to any other word. Some of these were specially created for particular branches of science (for instance, those used in the IUPAC[38] nomenclature system of organic chemistry – see **Chapter 11**); they are far too numerous or specialised to include here.

The most important biological suffixes are collected in the list below.

-*ase*: added to a stem word to indicate an enzyme (see -in);

☐ *amylase* (Latin *amylum*: starch) *noun,* an enzyme converting starch into sugar (a process which is very important for digestion – that is why its other name is *diastase* – Greek *dia*: aspect of thoroughness, completeness; Greek *stasis*: placing; Greek *diastasis*: separation).

-*cyte*: this suffix is used to indicate a cell type (Greek *kutos*: a container, a vessel, something hollow);

☐ *oocyte* (Greek *oon*: egg; *-cyte*: cell type) *noun,* egg-cell
☐ *histocyte:* (Greek *histion,* dimunitive of *histos*: something woven or net, tissue; *-cyte*: type of cell) *noun,* connective tissue cell.

-*gen* = -*genic* = -*genesis* = -*genous*: added to a word to indicate that the concept has something to do with the production or making of something (Greek suffix -*genes,* which is related to *gennan*: to produce);

☐ *antigen* (see above)
☐ *pathogen* (Greek *pathos*: suffering; *-gen*: producing) agent causing disease.

-*ia*: Greek ending without a special meaning, but used scientifically to give an indication that the concept refers to a condition of something;

☐ *diplopia* (Greek *diplous*: double; Greek *ops*: eye; *-ia*: condition of) *noun,* double vision
☐ *dysplasia* (Greek *dys-*: refers to something being wrong or bad; Greek *plassein*: to form to shape, to mould; Greek *plasis*: formation; *-ia*: condition of) *noun,* the condition of abnormal tissue formation.

-*in*: added to a stem to indicate an enzyme (see -ase);

☐ *pepsin* (Greek *pepsis*: digestion; *-in*: enzyme) *noun,* enzyme in weakly acid gastric juice which converts proteins into peptones

38 IUPAC is the acronym used for *The International Union of Pure and Applied Chemistry,* which is the international organisation which sets up the nomenclature of chemistry.

☐ *trypsin* (Greek *trypsis*: friction, from the Greek verb *tribo*: to rub; *-in*: enzyme) *noun,* enzyme occurring in the gastric juices of the pancreas (the name is related to rubbing or friction, because the enzyme was first obtained by rubbing the pancreas with glycerin!).

***-itic*: added to a word to indicate that the concept is an adjective referring to 'of' or 'in' (something);**

☐ *nephritic* (Late Latin abbreviation from the Greek *nephriticos*: of or in the kidneys) *adjective,* of or in the kidneys.

***-itis*: added to the stem of a word to indicate that it refers to the condition which is known as inflammation of an organ; the ending also occurs in some Greek feminine adjectives;**

☐ *nephritis* (Greek *nephros*: kidney; *-itis*: indicates inflammation of) *noun,* inflammation of the kidneys.

The names of people, places and habitats

The **names of people** – mostly those who discovered an organ, an organism or a theory, etc. – occur very often in biological nomenclature.

This is especially the case for the official names of plants and animals; in this case, there are definite rules which the international organisations have agreed upon and which govern the transliteration of these names into their Latin equivalents (some of which look rather funny and unfamiliar). The main rules are:

☐ when the name is already in Latin or Greek, then the genitive should be used, indicating *of*:

   – *alexandri* from Alexander
   – *francisci* from Franciscus or Francis.

☐ when a non-Latin or -Greek name of a person ends in a vowel, an *-i* is added, indicating *of*:

   – *glazioui* from the name Glaziou.

☐ when a non-Latin or -Greek name ends in a consonant, *-ii* is added at the end, indicating *of*:

   – *jamesii* from James
   – *macfadyenii* from the Scottish McFadyen, MacFadyen or M'Fadyen
   – *obrienii* from the Irish O'Brien
   – *leclercii* from the French Le Clerc
   – *vonhausenii* from the German Von Hausen.

☐ when the name functions as an adjective, then *-anum* or *-ana* are added:

   – *Geranium robertianum* from Robert(us) with genitive *Roberti*
   – *Verbena hasslerana* from Hassler.

The **geographical names** used on biological Latin names are rather difficult to classify, since they come from both ancient, medieval and modern place names; sometimes new Latin equivalents are simply constructed for places and localities the Romans never even heard of.

The best way to deal with classical and medieval names is to look them up in a very good dictionary, where such words as *Oxonia* (Oxford), *Lugdunensis* (Lyon), etc. will be found. Many Latin nouns are used to compound new names for regions; the most important are, for instance:

- [ ] *ager*: district
- [ ] *comitatus*: country
- [ ] *convallis*: a deep and enclosed valley
- [ ] *desertum*: desert
- [ ] *finis*: border, boundary
- [ ] *flumen, fluvius*: river
- [ ] *jugum*: mountain ridge
- [ ] *mons*: mountain
- [ ] *oppidum*: town
- [ ] *planities*: plain, steppe.

There are many more such words, but they are mostly recognisable from their English descendants, such as *paroechia* (parish), *lacus* (lake), *insula* (island), *districtus* (district), *provincia* (province), *promontorium* (promontory, headland or cape), and *vallis* (valley). These words are often used in descriptions, for instance, we say that a certain plant is found in *ager Lugdunensis,* that is, in the district of Lyon in France. The book by Stearn[39] contains a rather extensive list of Latinised geographical place names; the names in the list range from the obvious to the rather grotesque.

The cases where no suitable Latin or Greek names can be found

It is becoming increasingly difficult to find new names for *species* and *genera* which are subjected to reclassification or sub-classification – all the suitable Latin and Greek words, prefixes and suffixes may already have been used for similar species. This was already a problem for Linnaeus in the eighteenth century. The modern rules in this case are rather easy to remember:

- [ ] **invent a name** that looks Latin or Greek, but which does not really have to be quite so descriptive or accurate. It must be easy to pronounce;
  - the name can have a foreign origin, such as Japanese, Malayan, Indian, Arabic, etc.
    - **Arabic**: the plants called *Alchemilla, Taraxacum,* etc.
    - **French**: the plant called *Poncirum*

---

39 W T Stearn, *Botanical Latin*, Nelson, London, 1966. See also note 3.

- **German**: the plant called *Rorippa*
- **Indian**: the plants called *Madhuca, Maninkara, Vanda*
- **Japanese**: the plants called *Nandina, Sasa, Kirengeshoma,* etc.
- **Malaysian**: the plants called *Angraecum,* etc.
- **Persian**: the plant called *Jasminum*
- **North-American Indian**: the plant called *Camassia.*

☐ **make an anagram**[40] of the name of a similar species, keeping in mind that the word (which now has no real meaning since the original letters are now jumbled) must be easy to pronounce; some examples of anagrams are:

- *Sibara,* one of the *Cruciferae,* is an anagram formed from the name *Arabis* by totally inverting the word order
- *Diflugossia,* is an anagram formed from *Goldfussia* (which in turn, is derived from the German surname *Goldfuss*)
- *Lachemilla* (one of the roses) is an anagram formed from the related species called *Alchemilla* (which is of Arabic origin)
- *Mahernia* is an anagram formed from *Hermannia* by *Linnaeus,* judiciously leaving out an *n* to improve the pronunciation.

Some names even gave rise to more than one anagram to name-related species. For instance, the name *Allium,* one of the *Amaryllidaceae* gave the anagrams *Miluda* and *Muilla.*

Measure-
ments
All measurements are in the metric system, but there are a number of adjectives in use with which relative sizes can be expressed. These are, for instance, *altus* (high or deep), *latus* (broad), *crassus* (thick). Some nouns are also in use, but they are very easily recognised from their English equivalents, such as *altitudo* (height or depth), *longitudo* (length), *latitudo* (width), and *crassitudo* (thickness).

☐ Characteristic Terms

The accurate description of the object under discussion in a standard terminology which is understood by everyone is also very important in biology. This need has caused the invention of a very large descriptive terminology in biology – which must be learnt by the student. Most people learn by experience, that is, the terminology is picked up as we study.

It is much easier to develop an overview of the subject if one has some knowledge of the *system of terminology* of the particular branch of science

---

40 An *anagram* is the transposition of some or all the letters in a word or phrase to form a new word or phrase.

which one is beginning to study, and an attempt is made below to set up such a system of terminology for botany as an example, following Stearn.[39] There are two main classes of descriptive terms, namely the class of terms which refers to individual parts of the plant (leaf, stalk, colour, etc.), and a second class called collective terms, since they refer to the parts of the plant with respect to one another. These words occur over and over again in biology, and it is worthwhile to learn them (and the others not given here) by heart. Many (if not most) of the words also form part of the zoological vocabulary.

The **descriptive terminology** of botany depends very much on a set of characteristic terms under two classes:

Class I: Characteristic individual terms

Some terms are absolute, that is, referring to:

(1) The *figure* itself of the object:

    A. General or solid form;

        ☐ these terms include the following: *conicus* (conical), *cylindricus* (cylindrical), *tubulosus* (tubular), *fistulosis* (fistulous, or having a narrow passage or duct), *lachrimiformis* (tear-shaped), *cochleatus* (twisted so as to look like the convolutions of a shell with a converging helical shape), *fusiformis* (spindle shaped, tapering at each end), *teres* (not angular, usually applied to the stem of plants which are rounded), *semiteres* (flat on one side, *terete* on the other), *trigonus* (three-cornered), *angulosus* (angular; there are several types of this shape), *carinatus* (keeled, like a ship's bottom), *canaliculatus* (having a channel, channelled) (see **Figure 9.5**), and many more.

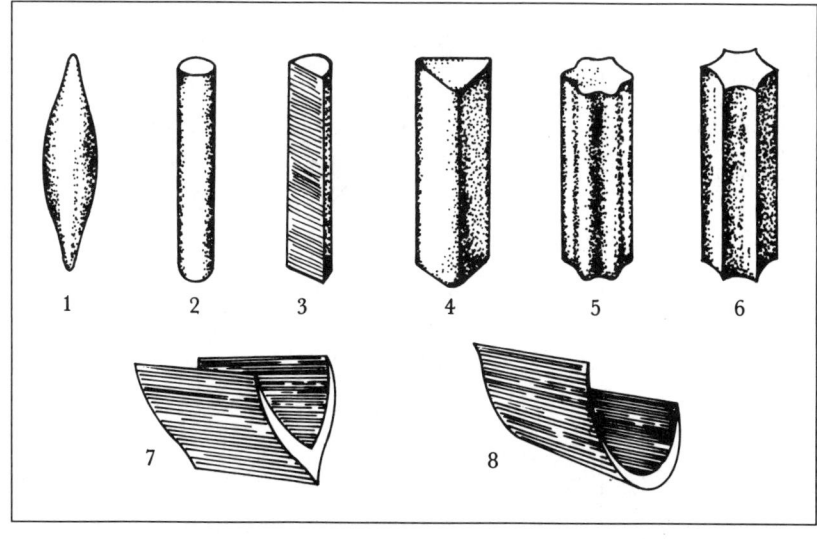

Figure 9.5
General or solid form. Sections of leaves and stems of plants.
These shapes are redrawn from *Stearn*[39], 321.
(1) *fusiformis;*
(2) *teres;*
(3) *semiteres;*
(4) *angulosus;*
(5) *angulosus;*
(6) *angulosus;*
(7) *carinatus;*
(8) *canaliculatus;*

B. Outlines and plane shapes;

☐ these terms include the following: *linearis* (linear), *fasciarius* (band-shaped, with the two opposite margins parallel), *oblongus* (oblong, elliptical and obtuse at each end), *ensiformis* or *gladiatus* (like the blade of a sword), *parabolicus* (parabolic), *rhombeus* (rhombic, shaped like a lozenge), *deltoides* (shaped like the triangular door of a tent, or a Greek capital letter delta), *cordiformis* (heart-shaped), *auriculatus* (shaped like the lobe of an ear), *lunatus* (shaped like the crescent of the moon), *reniformis* (shaped like a kidney), *sagittatus* (shaped like an arrow-head), *hastatus* (shaped like the point of a halbert), *panduratus* (shaped like a violin), *lyratus* (shaped like a lyre), *runcinatus* (runcinate, or with backward-pointing lobes), *undulatus* (wavy), *inaequalis* (unequal left and right sides), *dimidiatus* (halved, with one side almost non-existent) (see **Figure 9.6**), and many more.

Figure 9.6
Outlines of some of the shapes of leaves translated in the text.
Redrawn from *Stearn.*[39]
  (1) *ensiformis;*
  (2) *parabolicus;*
  (3) *rhombeus;*
  (4) *deltoides;*
  (5) *cordiformis;*
  (6) *auriculatus;*
  (7) *lunatus;*
  (8) *reniformis;*
  (9) *sagittatus;*
  (10) *hastatus;*
  (11) *panduratus;*
  (12) *lyratus;*
  (13) *runcinatus;*
  (14) *undulatus;*
  (15) *inaequalis;*
  (16) *dimidiatus.*

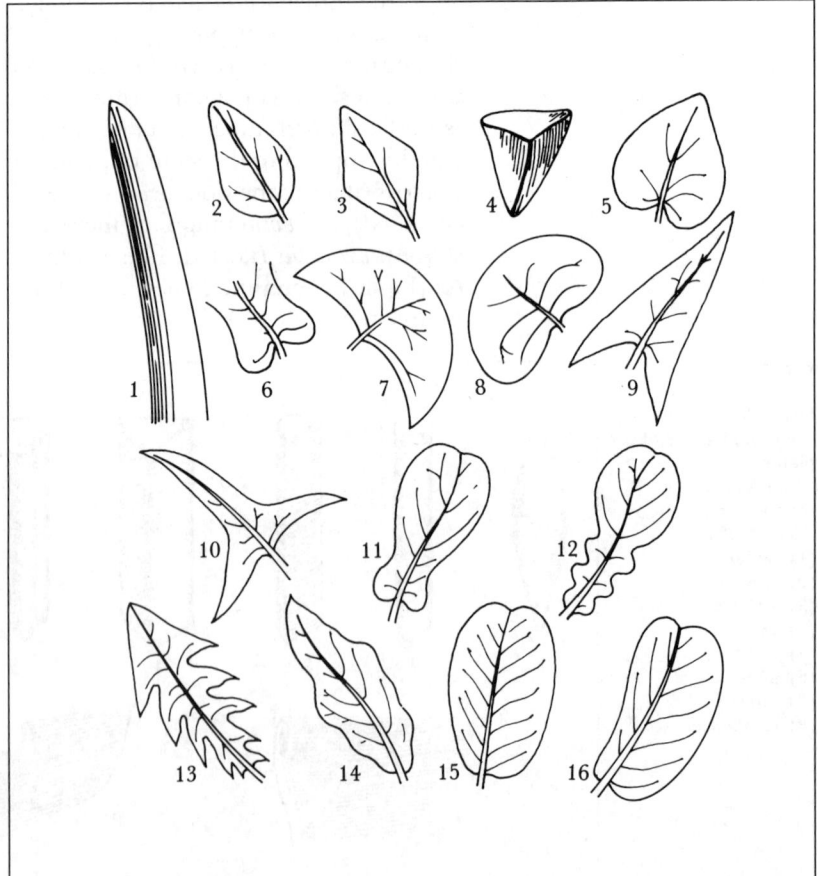

C. **The apex or the point of the object;**

☐ these terms include: *pungent* (pungent, terminating gradually in a hard sharp point), *setosus* (terminating gradually in a sharp point as in a bristle), *piliferus* (terminating in a very fine but weak point, hair-like), and many more.

D. **The base of the object;**

☐ these terms include: *cuneatus* (shaped like a wedge), *acutus* (acute, sharp) and a few more.

(2) **The *division* seen in the object:**

E. **The margin of the object;**

☐ these terms include: *serratus* (serrated, like the teeth of a saw), *dentatus* (toothed, having sharp teeth) *crispus* (curled), and a few more.

F. **Any incision seen in the rim or edge of the object such as a leaf;**

☐ these terms include: *lobatus* (lobed), *palmatus* (shaped like a palm leaf, but having five lobes), *digitatus* (fingered), and a few more.

G. **Any ramification or branching seen in a leaf;**

☐ these terms include: *simplex* (simple, almost not branched), *simplicissimus* (quite simple, having no branching at all), *pinnatus* (simple leaflets arranged on each side of a common petiole), and many more.

(3) **The *surface*:**

H. **Markings or evenness;**

☐ these words include: *annulatus* (ringed), *striatus* (striated or marked by longitudinal lines), *sulcatus* (furrowed or marked with longitudinal furrows or channels), and a few more.

I. **Hair-covering and superficial appendages or processes;**

☐ these terms include: *inermis* (no spikes, pines or prickles), *spinosus* (having sharp spines), *lanatus* (woolly), and some more.

J. **Polish or texture;**

☐ these terms include: *nitidus* (shining), *laevis* or *glaber* (smooth, but not shiny), *splendens* (glittering), *unctuosus* (greasy), *glaucus* (having the bloom occurring on the leaf of a cabbage), and a few more.

(4) Texture or substance;

☐ these terms include: *membranaceus* (like a membrane), *papyraceus* (like a sheet of paper), *oleaginosus* (fleshy, but filled with oil), *osseus* (bony), and a few more.

(5) Size;

☐ these terms include: *nanus* (dwarf), *pusillus* (very small), *elatus* (tall), *exaltatus* (lofty), and *giganteus* (gigantic).

(6) Duration or life span;

☐ these terms include: *annuus* (living only one year or growing season), *biennis* (living two growing seasons or two years), *polycarpicus* (bearing fruit a few times in one season), *perennis* (perennial, lasting for many years – some even for a few thousand years, like the California Redwoods), and some more.

(7) Colour;

☐ these terms include most of the Latin names for the colours, and they are mostly easily recognisable from their English equivalents.

(8) Variegation or variety;

(9) Veining;

☐ these terms include: *nervosus* (having several ribs), *palmiformis* (shaped like palm leaves), *venosus* (lateral veins variously divided), and a few more.

Relative terms

These terms refer to:

(10) Aestivation (manner of folding in the flower bud) and vernation (arrangement of leaves in the vegetative bud of the foliation).

(11) Direction, referring to such things as pointing towards the sun or floating upon the water.

(12) Insertion, referring to how something is attached to something else, such as sitting close on the body that supports it.

Class II Characteristic collective terms

(13) Arrangement, which refers to the relative position with respect to others and include such terms as opposite, adjacent, etc.

(14) Number, as discussed above.

☐ The Periods of Time in Biology and Palaeontology

We all agree how long a second, a minute, a week, a month, a year, a century or a millenium is. The second is exactly defined in terms of a physical standard (which is the frequency of a particular beam of light); the rest of the divisions of time follow from common consent: sixty seconds make a minute, sixty minutes an hour, etc. But there are other divisions of time which are not exactly defined. What is, for instance, meant by the term an *age*? The dictionary simply calls it a period of time. The word *aeon* or *eon* is then defined as a vast age, or eternity. These divisions are quite vague, yet we need them to systematise the vast periods of time in the history of the earth and thus for the description of the development of living organisms and their classification. We do this by allotting some periods of time to them and to other subdivisions of time which we need. And this is where the problem arises: not all of us agree on this. For instance, there are differences between the North American, South American, European and African names and the definitions of certain periods in the history of the earth. It is impossible to delve very deeply into these problems, and we refer the reader to Fenton and Fenton.[41] A brief summary of the main definitions is given below.

The time span we are talking about here is zero to 5000 million years, and we say that some (fossil) object has an age of (say) 100 million years.

> This time span of 5000 million years is subdivided into:
> - unequal *eons* (there are four of them)
>   - each eon is subdivided into one or more unequal *eras* (and in some cases into *sub-eras*)
>   - each era is subdivided into one or more unequal *periods*
>   - each period is further subdivided into one or more unequal *epochs*
>   - each epoch is divided into one or more *ages*
>   - each age is divided into one or more *phases*.

Each of these divisions of time has a name which is related to the main geologic conditions or to other identifying characteristics which

---

41 C N Fenton and M A Fenton (Revised and expanded by: P V Rich, T H Rich and M A Fenton), *The fossil book – A record of prehistoric life*, Doubleday, New York, London, Toronto, Sydney and Auckland, 1989.

occurred during that particular period of time. It is clear that we have again the familiar tree structure here which we also met when we discussed the systematisation of the living organisms of the earth.

The last 590 million years of the earth is called

☐ the **Phanerozoic eon** (Greek *phaneros*: visible; connective vowel -o-; Greek *zoikos*: of animals; suffix *-zoic* – the age in which the animals become visible, or in which they are seen for the first time).

The *phanerozoic* eon is divided into three *eras*:

☐ the *Cenozoic or Cainozoic era* (Greek *kainos*: new; connective vowel: -o-; Greek *zoikos*: animal – meaning that it had new animals which we now find as fossils); this era lasted from 65 million years ago up to now

– this era has two *sub-eras,* namely, *tertiary* (Latin *tertius*: third) lasting from 65 to 2 million years ago, and the *quaternary* (Latin *quaterni,* distributive of *quattuor*: four) which lasted from 2 million years ago up to now

– the tertiary divides up into the *Palaeogene* (65 - 25 millon years ago) and the *Neogene* (Greek *neos*: new) from 25 – 2 million years, and

– the quaternary goes over into the *Pleistogene* (Greek *pleistos*: most) for the last 2 million years

☐ the *Mesozoic era* (Greek *mesos*: middle; connective vowel: -o-; Greek *zoikos*: animal – the middle period of the animals), lasting from 248 million years ago till 65 million years ago

☐ the *Palaeozoic era* (Greek *palaios*: ancient; Greek *zoion*: animal – the most ancient period) which lasted from 590 to 248 million years ago.

Assignment 1

Use the above information to set up a time tree for the geological times, showing the number of years very clearly (Hint: use a non-linear time scale!). Use any encyclopedia to enlarge the tree to include the further subdivisions of periods and epochs; look up the meanings of the names which you do not know in the lists above, or in a good dictionary.

☐ Returning to the Classification System

We can now return to the binomial classification scheme which was discussed above. Let us reverse the order and let us start our discussion with the *Kingdoms,* instead of with the species. There are five Kingdoms:

☐ the **Prokaryota** (Greek *pro*: before; Greek *karyon*: seed) or the **Monera** (Greek *moneres*: single) referring to the bacteria and the blue-green algae – the simplest organisms.

☐ the **Protista** or the **Protoctista** (Greek *protista,* neutral plural superlative of *protos*: first) referring to those organisms which do not fit well anywhere else (this Kingdom is almost defined by the principle of exclusion!) and it contains things like the marine *diatoms* (Greek *dia*: through, apart, asunder; Greek *tomos*: from the verb temno: to cut), the slime moulds, etc.

☐ the **Fungi** (Latin *fungus*: a mushroom – probably from the Greek *sphongos*: sponge) containing the mushrooms, moulds and the lichens.

☐ the **Animalia** (Latin *anima*: breath, soul) refers to the animals with or without backbones, such as the molluscs, the mammals, etc.

☐ the **Plantae** (Latin: *planta*: plant) referring to all the plants, whether they contain chlorophyll or not.

The Kingdoms of the living organisms are schematically depicted in **Figure 9.7**.[42] Even though this is not a textbook of biology, it is very worth-while to make a close study of this figure – without having to go into the details of the subject itself. This family tree of the living organisms, as depicted in **Figure 9.7**, deserves to be treated as much more than just a pretty picture to illuminate this book. This rather simple-looking figure lies at the linguistic and the scientific heart of biology. It should be subjected to a very close study in order to analyse the scientific model of nature which the biologist has built up to explain nature; for instance, one can think about:

☐ the principle of branching and the type of questions which must be asked about the organisms to determine whether one goes to the left, to the right, or upwards in the branching diagram

☐ the principle of the setting up of the sets of properties in order to give a name to an organism

☐ the topology of the Kingdoms: the *Monera* at the bottom, the *Protista* just above it and touching the Kingdoms of the *Plantae,* the *Fungi* and the *Animalia* at their undersides

---

42 There is no general agreement between biologists about the final details of such a tree-like structure, although they do agree on the broad outlines. For instance, Margulis and Schwartz[25] display a somewhat different figure, but the same questions as above can be asked about their figure. In addition, one can ask: *why do the two figures differ?* The answer to that is simple: *we need more research in order to resolve these differences!* There is still much research to be done in all branches of biology!

☐ the fact that the Kingdoms of the *Plantae*, the *Fungi* and the *Animalia* do not touch, while there are other common boundaries
☐ the implication which this branching diagram has for the concept of development over a (very long) period of time
☐ the implications of the fact that the Kingdom *Plantae* is directly linked to the Kingdom *Monera* without any branchings in the Kingdom *Protoctista*.

Hint: Think of properties in terms of sets and subsets!

☐ the implications of the fact that the *Chlorophyta* (the grass green algae) are linked to *two* Kingdoms, namely the *Monera* and the *Protoctista*.

Note: This is the one instance where there seems to be *convergence* in the tree – all other relationships are *divergent;* this needs some careful thought.

Hint: Think of properties in terms of sets and subsets!

☐ the implications of the fact that the *Ginkgos* in the Kingdom *Plantae* are near-neighbours of the *fungi* in the Kingdom *Fungi,* or the *Angiospermopsida* in the Kingdom *Plantae* and the *Jawless fish* in the Kingdom *Animalia* (Question: are they converging together?).

Hint: In topological branching diagrams the only important things are the nodes and the links!

Hint: Think of properties in terms of sets and subsets!

☐ the implications of the fact that a *horse* shares the same branch with a *woman and her baby* (Question: what do we have in common?).

Hint: Think of properties in terms of sets and subsets!

One should also:

☐ *identify five Kingdoms and their relationships and the typical Phyla in each one;* for instance, the meaning of the name *Monera* has implications for the theory of the time-development of organisms from a period where there was nothing alive on earth.
☐ *learn the names of the Phyla and their meanings so that they can be under-stood and remembered.*

Figure 9.7
The family tree of presently living organisms, showing the Kingdoms, their interrelationships, and their *Phyla*. Reproduced with permission from *Fenton and Fenton*.[41]

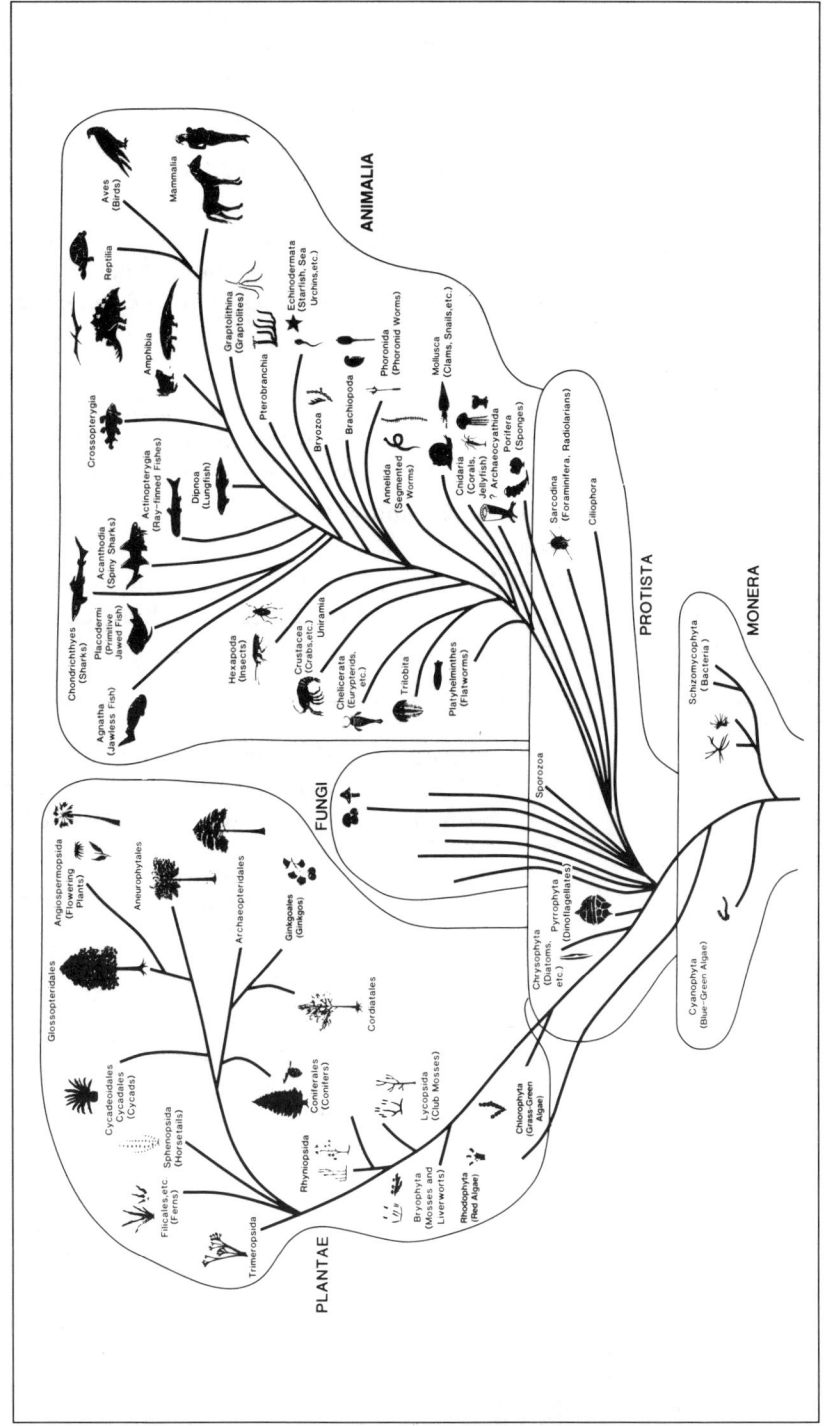

If we now think of the horse and the woman with her baby depicted on the same branch in the Kingdom *Animalia,* then it becomes clear that they share the same subset of characteristics which we use to identify and name the *Class Mammalia.* These two organisms are, however, so different, each having their own individual properties, in addition to the *Class* properties, that it follows that there must be more branching points to follow. Adding more properties to that of the *Order,* we obtain those of the *Family.* And finally, we obtain the *Genus* and the *species* of the horse and the human being. We have now reproduced the argument above, coming from the opposite direction.

Assignment 2

Working backwards from the horse and the woman and the baby which belong to the *Class Mammalia,* try to determine the branching point which leads to:

the *Subphylum: Vertebrata;* and that of
the *Phylum: Chordata;*

You need very little biology for this – use your common sense. The word *Chordata* refers to the fact that the animal has a spinal chord encased in a protective structure. The word *Vertebrata* means that the structure enclosing the spinal chord is calcified, that is, of bone.

The classifications of the horse and the human being are displayed below in **Table 9.1.** It is clear that they share many properties as the common classification down to the level of the *Class* shows.

Table 9.1
The classification of the horse and the human being.

Taxonomic level	The Human Being	Horse
Kingdom	Animalia	Animalia
Phylum	Chordata	Chordata
Subphylum	Vertebrata	Vertebrata
Class	Mammalia	Mammalia
Order	Primates	Perissodactyla
Family	Hominoidea	Equidae
Genus	Homo	Equus
species	sapiens	caballis

Assignment 3

Draw the branching diagram of the horse and the human being, using the information from **Table 9.1.**

□ The Terminology of Size

**Intro-
duction**

The determination of size – both relative size, as well as absolute size – plays an important role in biology: we simply have to know how big things are. An extensive terminology has been developed to deal with aspects of measurement of sizes, such as:

□ the terminology of *relative sizes* (which we have already dealt with above)

□ the terminology related to the use of the *naked eye* with which we can see objects in the range 10^{-3} meter (= 1 millimeter) to 10 meter with sufficient detail for descriptive processes (this will not be pursued here)

□ the terminology relating to the simple *magnifying glass* which is held in the hand and with which we can see objects in the range 10^{-4} meter (0.1 millimeter) to 10^{-2} meter (10 millimeters) with sufficient detail for descriptive processes

□ the terminology relating to the *compound optical microscope,* sometimes called the *light microscope,* with which we can clearly see objects in the range of 10^{-6} meter (0.001 millimeter) to 10^{-4} meter (0.1 millimeter)

□ the terminology relating to the *scanning electron microscope,* commonly abbreviated *SEM* (which uses an electron beam instead of light and electron lenses instead of glass or quartz lenses), with which we can see objects from about 10^{-8} meter (0.00001 millimeter) in size up to about 10^{-2} meter (10 millimeter or 1 centimeter)

□ the terminology relating to the *transmission electron microscope* (commonly abbreviated *TEM*) with which we can see very thin and very small objects in size from about 10^{-9} meter (0.000001 millimeter) to 10^{-5} meter (0.01 millimeter).

Since it is unlikely that a student will have access to the last two instruments, we will not discuss their terminology any further; pictures produced by such instruments are similar to those of conventional microscopes, and are to be treated as such.

The word *microscope* is derived from the Greek *micros*: small (modern Latin *-scopium,* from the Greek *skopeo*: to see), that is, it is an instrument which is used to see small objects. We use the term to *magnify,* which means: *to increase the apparent size of an object.* The microscope or the magnifying glass does not really increase the size of the object; they merely present us with a *virtual image* which the eye perceives as being bigger than the object itself. Even though the eye cannot directly see the virtual image of a *SEM* or of a *TEM* (we eventually see only the photographs, such as that of **Figure 9.1** on page 132), their terminology is based upon that of the optical microscope.

There are some further techniques in the life sciences which are used to view parts of organs or bodies, but which are only mentioned here, since the beginner does not usually see pictures of objects which were taken using them; for instance:

☐ the terminology relating to the *nuclear magnetic resonance technique* in which a computer processed optical image (usually called a *CAT Scan*) of a section through the object is presented on a TV screen; this image can be manipulated and enlarged almost at will.

Each picture or sketch of an object or organism must be accompanied by a statement of size. It was usual in the older literature just to quote the *magnification* of the microscope used, but modern texts usually add a *size bar* at the side or at the bottom of the drawing or photograph as seen in **Figure 9.1**.

Termin-ology relating to the magnifying glass

The magnifying glass is the simplest of the instruments used to magnify an object; its basic principles are shown in **Figure 9.8**. If the object we wish to see is situated on the focal point of the lens then the *magnifying power* is given by the expression v/f, where f is the *focal distance of* the lens, and v the *distance of distinct vision*. This distance of distinct vision varies with age and from person to person, but a working average is about 25 cm. The usual magnifying power of a single lens is of the order $4x$ to $8x$, where the x is to be read as 'times'; larger values can be found (up to about $20x$), but these usually introduce considerable distortion outside the centre of the lens.

Figure 9.8
The basic principle of the common magnifying glass

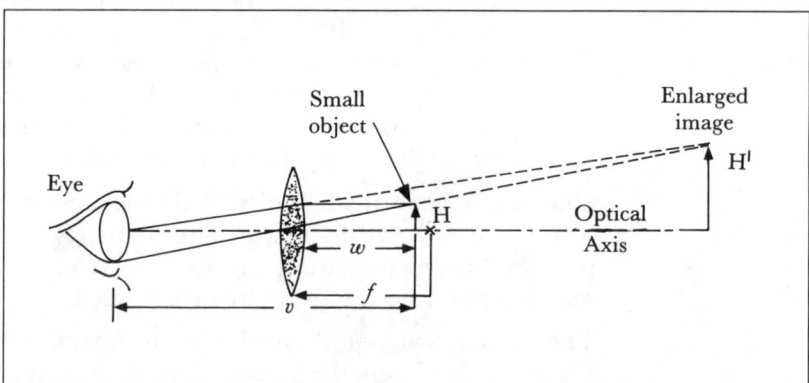

The optical microscope

The optical microscope makes use of two sets of compound lenses called

☐ the *objective,* situated near the object which is to be magnified with a focal length f and initial magnification v/f

☐ the *ocular* or *eyepiece* into which the person is looking, with focal length F and magnification a/F, where a is the distance between the two focal points which is called the optical tube length.

The total magnification is the product of the magnifications of the objective and the ocular, and it can be as high as 2000*x*. The principle is illustrated in **Figure 9.9**.

Figure 9.9
The optical principle of the light microscope.

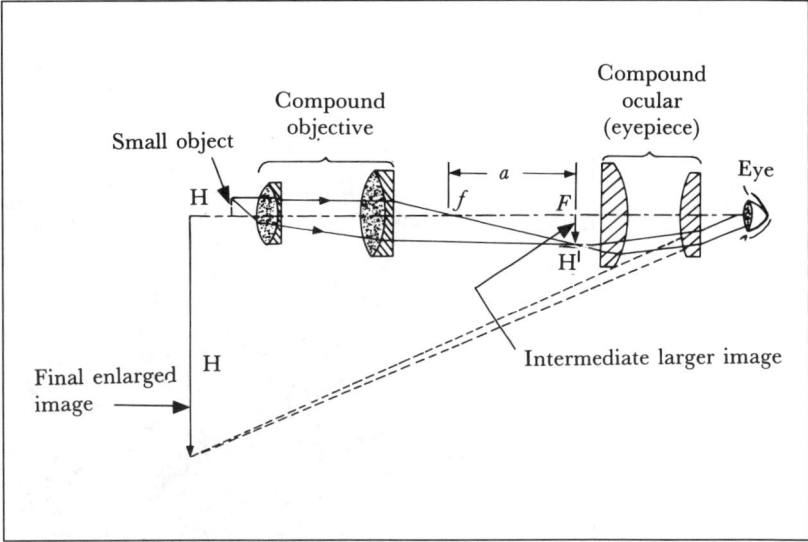

Figure 9.10
Names connected with the microscope

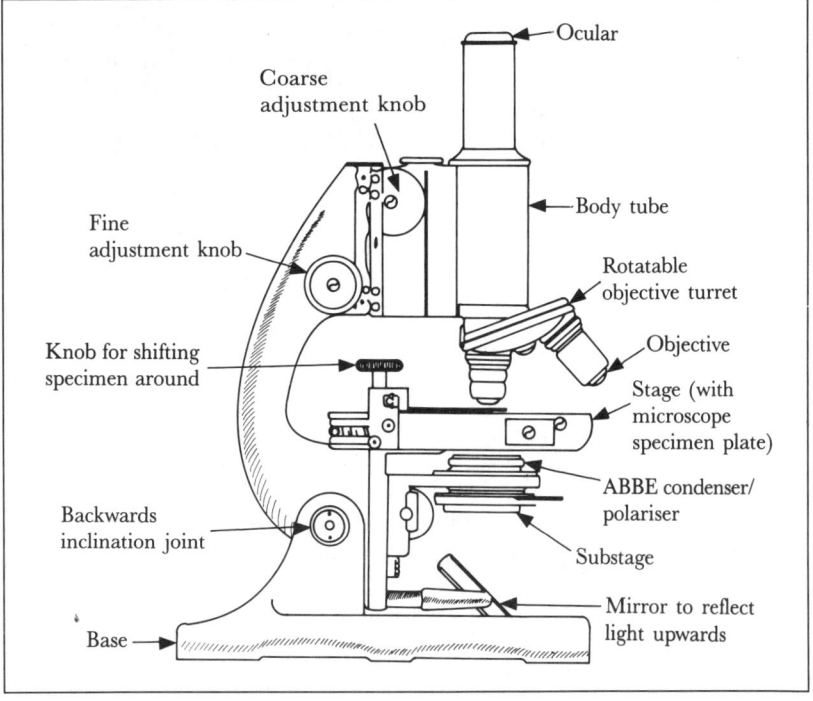

Assignment 4

Try to determine the meanings of the names in **Figure 9.11** of a unicellular organism. (Used with permission from Margulis and Schwartz[25])

Figure 9.11
A generalized poly-cystine actinopod in cross section. Thick black lines = skeleton; stippled areas = cytoplasm; N = nucleus. [Drawing by L. Meszoly, information from Georges Merinfeld.]

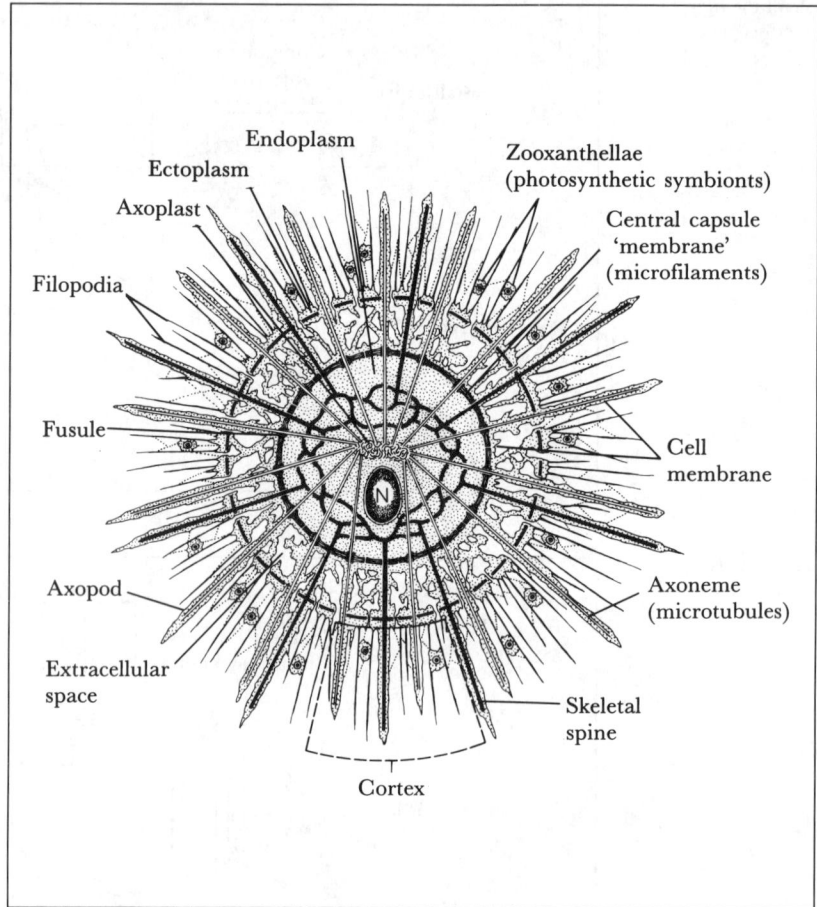

Endoplasm

Ectoplasm

Axoplast

Zooxanthellae (photosynthetic symbionts)

Central capsule 'membrane' (microfilaments)

Filopodia

Fusule

Cell membrane

Axopod

Axoneme (microtubules)

Extracellular space

Skeletal spine

Cortex

Assignment Answer 1

See **Figure 9.12**, which is reproduced in a somewhat changed form from Fenton and Fenton[41] and is used with permission. The latest (1989) geological time scale differs in minor respects from this one.

☐ Concluding Remarks

Some aspects of the language used in the biological sciences were discussed in this chapter. We gave special attention to *Biological Latin*, emphasising the fact that almost every biological noun, verb and adjective has a meaning which can be deduced from its roots. To enable the reader to fend for himself or herself, we showed how the biological terms which are derived from Latin and Greek can be broken up into their constituent parts, each of which combine with the other part(s) to give the full meaning of a word. Many examples are given, illustrating the use of those words (nouns, verbs, adjectives, prefixes and suffixes) which are often used in the construction of biological terms. It was stressed that the meanings of many biological terms can be found by the judicious use of most good dictionaries which contain etymologies; the reader is urged never to learn a term without first finding out its entymology.

We have also discussed some elementary aspects of the process of giving names to biological species, such as plants or animals. This binomial system of nomenclature forms one of the cornerstones of biological terminology and the reader is urged to make a thorough study of the subject, using textbooks of Botany or Zoology.

Figure 9.12 (Reproduced with permission from *Fenton and Fenton*.[41])

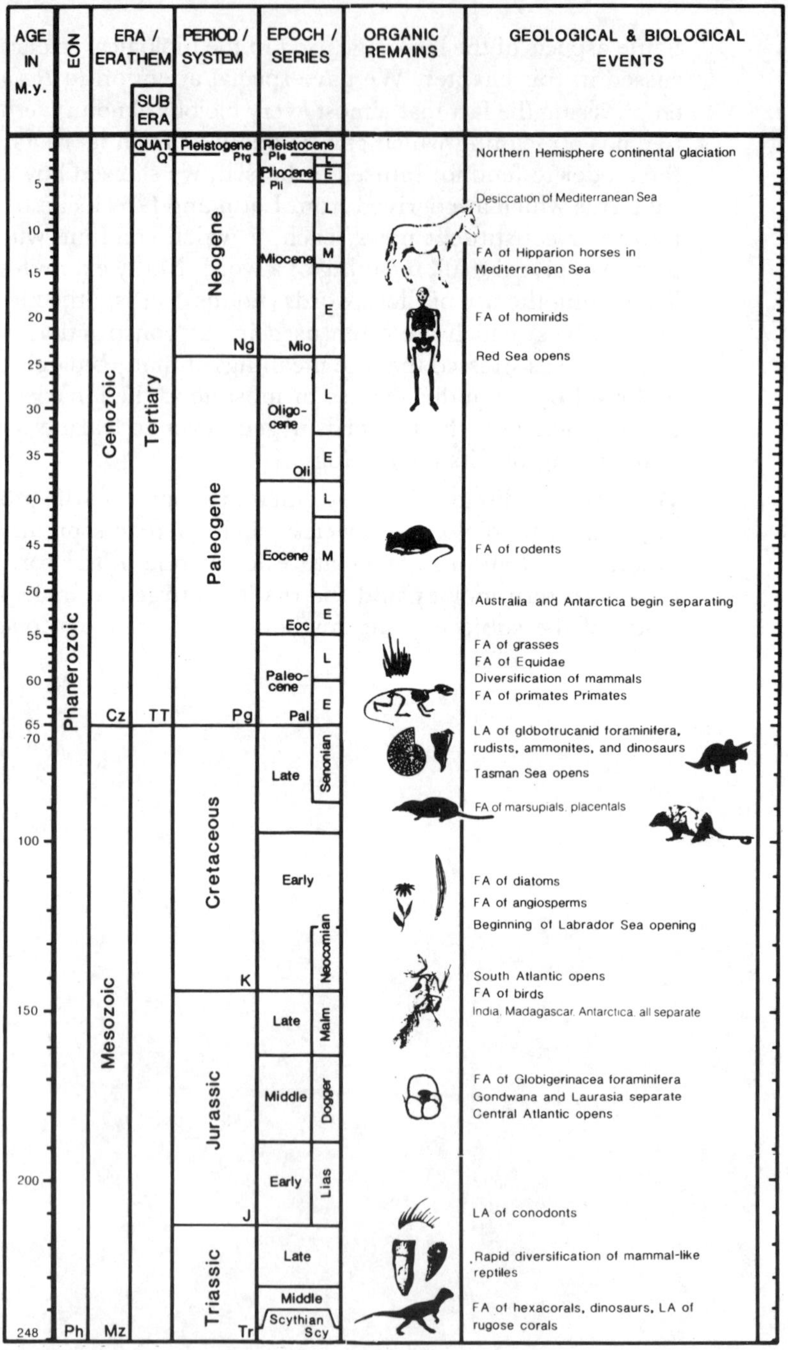

FA: First Appearance;   LA: Last Appearance

Figure 9.12 *(continued)*

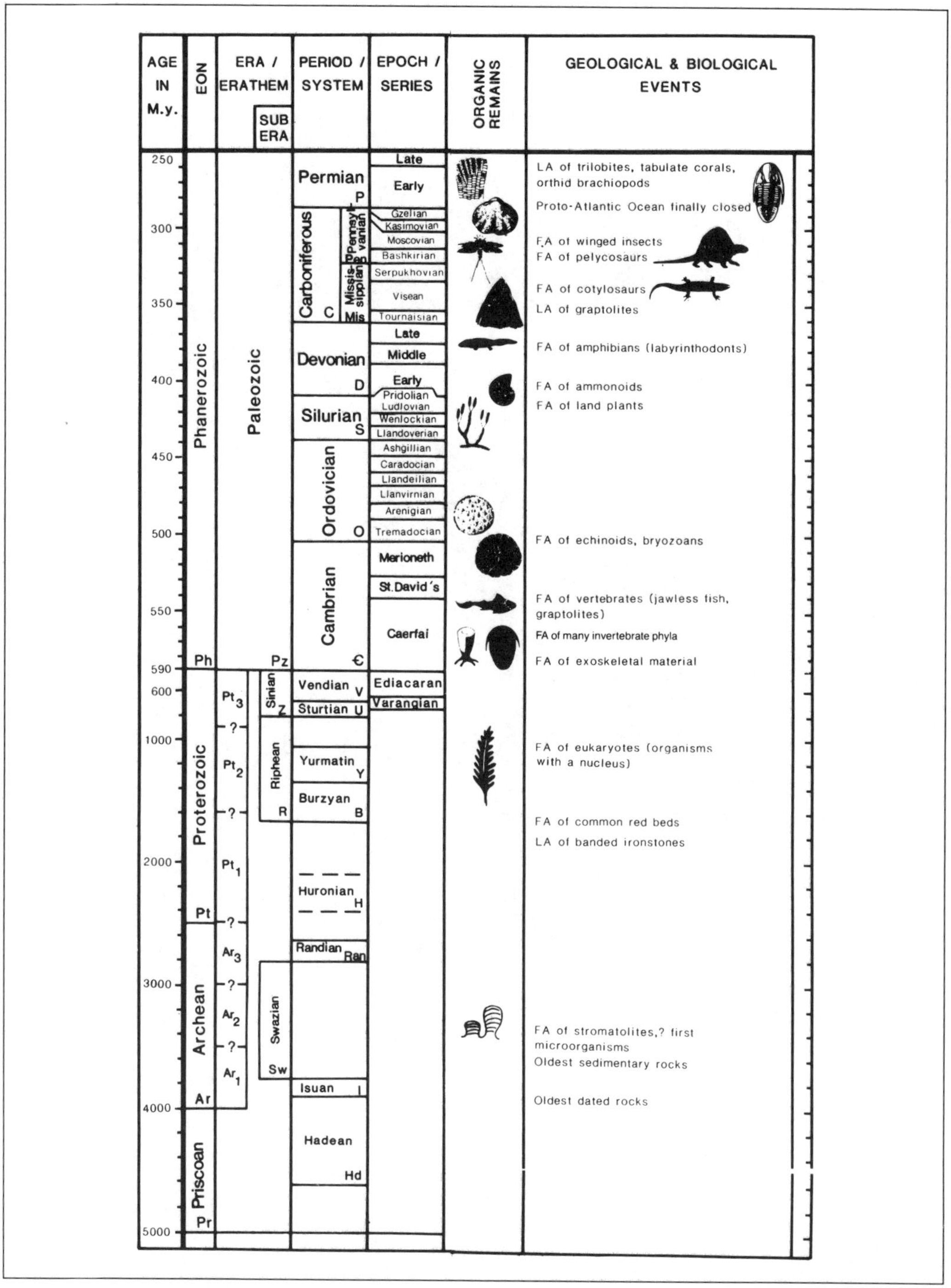

AGE IN M.y.	EON	ERA / ERATHEM	SUB ERA	PERIOD / SYSTEM	EPOCH / SERIES	ORGANIC REMAINS	GEOLOGICAL & BIOLOGICAL EVENTS
250	Phanerozoic	Paleozoic		Permian P	Late		LA of trilobites, tabulate corals, orthid brachiopods
					Early		Proto-Atlantic Ocean finally closed
300			Pennsylvanian	Carboniferous C	Gzelian		
					Kasimovian		FA of winged insects
					Moscovian		FA of pelycosaurs
					Bashkirian		
			Mississippian Mis		Serpukhovian		FA of cotylosaurs
350					Visean		LA of graptolites
					Tournaisian		
				Devonian D	Late		FA of amphibians (labyrinthodonts)
					Middle		
400					Early		FA of ammonoids
				Silurian S	Pridolian		FA of land plants
					Ludlovian		
					Wenlockian		
					Llandoverian		
450				Ordovician O	Ashgillian		
					Caradocian		
					Llandeilian		
					Llanvirnian		
					Arenigian		
500					Tremadocian		FA of echinoids, bryozoans
				Cambrian €	Merioneth		
					St. David's		FA of vertebrates (jawless fish, graptolites)
550					Caerfai		FA of many invertebrate phyla
590	Ph	Pz	Pz				FA of exoskeletal material
600		Proterozoic Pt₃	Sinian Z	Vendian V	Ediacaran		
				Sturtian U	Varangian		
1000		Pt₂	Riphean R	Yurmatin Y			FA of eukaryotes (organisms with a nucleus)
				Burzyan B			
							FA of common red beds
							LA of banded ironstones
2000		Pt₁		Huronian H			
	Pt						
		Ar₃		Randian Ran			
3000		Ar₂	Swazian Sw				FA of stromatolites, ? first microorganisms
	Archean						Oldest sedimentary rocks
		Ar₁		Isuan I			
4000	Ar						Oldest dated rocks
	Priscoan Pr			Hadean Hd			
5000							

10

An Introduction to the Language of Mathematics

In the opinion of many laymen mathematics is today already a dead science: after having reached an unusually high degree of development, it has become petrified in rigid perfection. This is an entirely erroneous view of the situation; there are but few domains of scientific research which are passing through a phase of such intense development at present as mathematics. Moreover, this development is extraordinarily manifold: mathematics is expanding its domain in all possible directions, it is growing in height, in width, and in depth. It is growing in height, since, on the soil of its old theories which look back upon hundreds if not thousands of years of development, new problems appear again and again, and ever more perfect results are being achieved. It is growing in width, since its methods permeate other branches of sciences, while its own domain of investigation embraces increasingly more comprehensive ranges of phenomena and ever new theories are being included in the large circle of mathematical disciplines. And finally, it is growing in depth, since its foundations become more and more firmly established, its methods perfected, and its principles stabilized.

Alfred Tarski[1]

☐ Introduction

Mathematics is expressed in a complex, but very precise language which is quite different from that of normal usage or even from that which is used in other branches of science. This language has ordinary language and logic as its basis, but superimposed on this is an array of precisely-defined conventions, notations, symbols, graphical and abstract representations, as well as a whole range of mathematical actions or operations described by special verbs, adjectives, adverbs and verb phrases which are also very precisely defined.

1 Tarski A. *Introduction to logic and the methodology of deductive sciences,* Oxford University Press, Oxford and New York, Third Edition, 1965.

Mathematics is an abstract science and it must always be remembered that it needs to be communicated between people. The reader of a mathematical text must, therefore, realise that its author is speaking directly to him or her in this very precise language. The author, furthermore, assumes that the reader *knows and understands the mathematical language, as well as all the notations, conventions, definitions, symbols, actions and verbs, etc* which are going to feature in the discussion. This means that the reader must be well prepared for the exposure to the new material which will be introduced. This holds both for an elementary book intended for educational purposes, as well as for a research paper in which a mathematician communicates research results to other mathematicians.

This mathematical language has developed over a period of more than three thousand years – slowly and laboriously from ancient times up to the Middle Ages, faster during the Renaissance, exploding boldly during the 16th to the 20th centuries, and is still developing today at a rapid pace.

It is not the purpose of this chapter to teach the subject of mathematics, nor to expose the reader to profound philosophical notions of mathematics and linguistics and to advanced aspects of the application of linguistics to the language of mathematics. To do so would confuse beginners, especially if they are not completely familiar with the English language, or if they are almost frightened by the mathematical texts with which they are coming into contact. The attention of the beginner (who is not completely familiar with the English language) is, therefore, drawn to some of the aspects of the usage of language in mathematics which are necessary in order to read and understand written mathematical texts. Inevitably, however, the illustrative examples in the text involve (elementary) mathematics, but *no attempt is made to weave them into one coherent mathematical system;* some terms even occur more than once. We emphasise the following:

> The reader is expected to concentrate on the **language** and the **flow of argument,** and not to try to understand the mathematical examples.

These examples, coming from many branches of mathematics, are chosen in such a way that they do not make great demands upon the mathematical knowledge of the reader.

The following type of question about this language of mathematics may be asked:

☐ *Why do we need this very precise and subtle language of mathematics at all?*

☐ *Is it not just as easy to do mathematics using ordinary English?*

Before answering these questions, it is instructive to read a quotation from a fourteenth century Italian mathematical text which uses the so-called *rhetorical style*. This style predates the modern way of writing algebraic equations, as well as the use of mathematical symbols (except those representing natural numbers and fractions). The author intends to prove that multiplying a negative number by a negative number results in a positive number,[2] and the details of his proof are of importance. It should be noted that the proof which is offered is not a proof in the modern sense of the word: he appeals to experience to extrapolate from the one specific case (which he shows to be true) to the general case (which he assumes to be true). The translation reads as follows:

> **And multiplying minus times minus makes plus.** *If you would prove it, do it thus: you must know that multiplying 3 and ³⁄₄ times 3 and ³⁄₄ will be the same thing as 4 minus ¹⁄₄ times 4 minus ¹⁄₄. That is, multiplying 3 and ³⁄₄ times 3 and ³⁄₄ makes 14 and ¹⁄₁₆; as does multiplying 4 minus ¹⁄₄ times 4 minus ¹⁄₄. And to be clearer multiply per chasella[3] saying 4 times 4 makes 16, now multiply across and say 4 times minus one quarter makes minus 4 quarters, that are one integer, and 4 times minus one quarter makes minus one, you have minus 2, take this from 16 and it leaves 14. Now minus ¹⁄₄ times minus ¹⁄₄ makes ¹⁄₁₆, that makes one as much as the other.*

Problem 1

Try to rewrite this fourteenth century rhetorical description which proves that $(-\frac{1}{4}) \times (-\frac{1}{4}) = +(\frac{1}{4} \times \frac{1}{4}) = (\frac{1}{16})$, using modern mathematics.[4]

Another example is the statement that *twice the product of any two consecutive numbers is equal to the sum of their squares less one*. This statement looks difficult to comprehend to the untrained eye and mind, but if it is translated into the symbolic mathematical language which we all know, it becomes very easy:

$$2(y + 1)y = (y + 1)^2 + y^2 - 1$$

---

2 Franci R and R L Rigatelli, Fourteenth century Italian algebra, in: Hay C (Editor), *Mathematics from Manuscript and Print 1300-1600*, Clarendon Press, Oxford, 1988.
3 **Per chasella** is just ordinary cross-multiplication where $(a + b)(c + d)$ gives $ac + bc + ad + bd$.
4 The solutions to the problems are to be found at the end of the chapter.

where $(y + 1)$ and y are the two consecutive numbers we are talking about;[5] when we multiply out both sides, the identity follows.

Let us reverse the procedure, using an example from Tarski[6] which should be carefully studied and then translated into written English:

For any numbers x and y

$$x^3 - y^3 = (x - y). \; (x^2 + xy + y^2).$$

In normal English this would look as follows:[7]

> The difference of the third powers of any two numbers x and y is equal to the product of the difference of these two numbers and a sum of three terms, the first of which is the square of the first number, the second the product of the two numbers, and the third the square of the second number.

Four important points emerge when mathematical statements such as these rhetorical statements are carefully studied:

(1) The *rhetorical style* is not clear; it is not easy to understand at a first reading; it actually calls for repeated reading and thinking and rethinking. In fact, one has the feeling that writing like this *ought to be avoided at all costs.*

   – Mathematics places great emphasis on the ability to write **clearly** and **concisely,** and above all, **unambiguously.** For this purpose the elaborate and precise language of mathematics was developed. Every student should make a special effort to understand this very austere and yet elegant language with which thoughts can be expressed in such a way that another mind can grasp them in an unambiguous fashion. The language of the mathematician in this respect is just the opposite to the language of the poet; yet, the most elegant and sublime thoughts can be expressed by both.

(2) The author of each of these mathematical statements assumes that the reader already has some knowledge of numbers and arithmetic. The reader is, furthermore, expected to apply this

---

5 This way of doing things is quite common in mathematics, and very many examples are found in which we generalise things by saying something like: Let the first number be y; then the next number is $(y + 1)$ and the *previous* number is $(y - 1)$. From these assumptions we get their products, sums, differences, ratios, etc. which help us to solve posed problems.

6 See footnote 1.

7 One of the great inventions of mathematics is the introduction of the signs for mathematical *operations.* For instance, the sign + which signifies *addition* is derived from the sign which the medieval scribe used to abbreviate the word 'and' in his text, namely the *ampersand* , &, which was hastily written in the form of a ' + ', with the right-hand and top sides still connected. The great improvement brought by the introduction of the little sign ' – ' to indicate a *negative number* (which was previously written *neg. a*) or *subtraction* (actually, *addition* of a negative number!), is easily evident.

knowledge in a practical way (to execute multiplications, etc) in order to understand a proof or to arrive at the required results.

- All mathematics is interrelated: what has gone before should always be remembered, since an author always assumes that the notation, verbs, definitions, etc are known to the reader from the preceding pages, or even from the studies of the previous year or years (even that which was learnt at school is always relevant in mathematics). Some of this general knowledge in mathematics will simply be used in advanced textbooks without ever specifically referring to it: the author assumes that the reader already knows it.

(3) The statement from 14th century Italy, furthermore, contains a number of instructions denoted by the imperative form of verbs such as *multiply* (or in the form multiplying) and *take* which are directed at the reader, instructing him to do something which the author thinks that the reader should be able to do by herself (that is, these instructions or operations were defined earlier in the book, or elsewhere).

- Imperatives such as these still occur in modern mathematics, and every writer confidently expects that the reader can and will execute them for herself in the correct way. Wherever they (or other similar instructions) occur, the flow of the argument demands that the reader executes them in order to supply the missing parts herself. Reading mathematics is thus an active occupation; pen and paper are needed for these activities (unless one is a very good and experienced mathematician). Mathematics is not learnt like a poem or the clauses in a historical treaty. The definitions, symbols, notations, etc of mathematics can be learnt in such a way that they are understood, but mathematics is done.

(4) The proof given in the medieval Italian quotation is actually just a recipe, since it involves only one single example, and is not general at all. This means that it does not prove that all conceivable negative numbers behave in this way when multiplied together.

- Mathematics places great emphasis upon the general and tends to stay away from the particular (unless a specific aspect or problem is needed as an illustration or as an exercise). It is for this purpose that the elaborate and precise system of symbolic notation was developed which enables the mathematician to deal with whole ranges of numbers, for instance, at the same time. This system of notation is very important and attention should be given to learn and to understand it.

Variable   One of the great discoveries of mathematics is that of the concept variable which allows it to deal with a whole range of similar things at
the same time; this is one of the great abstractions which the human
mind has achieved. An example, for instance, is the variable which
we'll call x, where x is any member of the set **R** of all real numbers.
It must always be remembered that eventually the name of a number will have to be substituted for the variable. It therefore follows
that a variable in mathematics is the placeholder of a noun, such as
the name of a number. This is emphasised by sometimes placing a
symbol such as a block ■, a diamond ◆, or any other defined symbol called a placeholder instead of the general variable (for instance, x):

identity   (■ + 1) (■ − 1) = ■² − 1

equation   2 ■ − 1 = 3.

☐ Some Aspects of the Nomenclature of Sets and Numbers

The concept of set is very basic in our everyday lives, as well as in
mathematics. A set is basically a collection of objects, and it really
does not matter whether this collection is of real objects (like stamps,
or a tea set) or of imaginary objects (like numbers, ghosts, or a specific
number of games won by a player in a tennis match to win a set).

It is important to note that a set should be well-defined in the sense
that when given any object it should be possible to decide whether
or not it belongs to the set under consideration. For example, the set
of the ten most important people in the world is not a mathematically well-defined set, since every person would come up with a set
of personal favourites.

What matters about a set is the fact that we ourselves can decide what
to put into any set we choose to describe, and also how many members (also called elements) it should have.[8] We do not physically put
the objects we choose into a little enclosure which we call by the name

---

8 In ordinary life we are used to the concept of the full set. For instance, the house-
wife has a *tea set* (which contains a specified set of utensils, like the tea pot, the
sugar bowl, the milk jug and 6 cups and saucers – all of which look alike by
having the same design and colouring). If the milk jug happens to be broken,
then the set is broken and there is actually no set left to display to the neighbour.
The philatelist collects a *set of stamps,* that is, all the stamps which were issued
by a postal authority on a certain day to commemorate an event; half a set of
stamps is no set at all. For the mathematician, half of the set of stamps will still
be a set, or the collection of the broken tea set, since any collection of objects is a set.

of set. A set in mathematics is a list of things, and this allows us to clearly distinguish between a set with one member called *x* and the real object it refers to. For instance, we cannot fetch the real object (the queen Cleopatra) to put into the set {*Cleopatra*}, which contains only one member, namely the object whose name is *Cleopatra*. Such a set with only one member is called a singleton. It should be noted that one distinguishes between, for example, the number 1 and the *set* {*1*}, which has the number 1 as its only element. This so-called *list notation* (also called *roster notation*) in which the individual members of a set are listed one by one between curly braces is convenient only for sets with very few members; it becomes cumbersome with larger numbers of members (the names in the complete New York telephone directory may form such a set!) and it cannot be comprehended at a glance. In any case, such a list cannot be constructed for a set with infinitely many members, although we can, for example, write the set of all even natural numbers as {*2, 4, 6, ...*}, where the row of dots indicate that we are supposed to continue the list to infinity. It is often more convenient to use the *set-builder notation* for sets which we'll discuss below, because it allows us to contract a whole list into a small collection of symbols, each having its own meaning.

Mathematicians have developed this set-builder notation to describe a set. The set with name *A* having members *x* satisfying some property called *P(x)* is written as follows, where the *curly brackets or braces* and the sign = form part of the definition:

$$A = \{x \mid P(x)\}.$$

The upright bar | is the symbol which represents and abbreviates the words: *such that ... holds*. The symbol *P(x)* is an abbreviation for some property *P* which is precisely-defined in order to specify the *x*'s in the set. The property *P*, as it were, picks out *all* the relevant *x*'s from all the other possible similar objects in the universe and lists them in the set which we call *A*.

It is perhaps important to pause a moment and to look at the meaning of the verb *to hold*. The verb *to hold* has a range of meanings in ordinary English, but in mathematics it means something like: *is true*. The definition of the set *A* then reads as follows:

> *The set A is the set of all x such that the property P(x) holds.*

The example of Cleopatra can then be written $A = \{x \mid x$ *is the name of a Queen of Egypt who loved both Mark Anthony and Caesar*} = {*x* | *x is Cleopatra*}.

It is seen that every symbol in the definition is important, and that every symbol is read, and has a meaning. Especially important is the insertion of the adjective *all* to describe the *x*. This is an exhaustive property: the definition rigorously includes all elements x which have the defining property $P(x)$ and excludes all others. More examples of sets will be found below.

What also matters in the case of a particular set is:

☐ whether we can clearly distinguish between the members of the collection or set (for instance, all golf balls look identical, but we can put an individualising production or other number upon each of them, thus distinguishing between them, as it were, giving a personal name to each) so that we can enumerate or list them for the description of the set containing certain golf balls which we have chosen to put into it; the set $\{a,\ a,\ b,\ c,\ d\}$ is not a set, but if we omit one of the a's we do obtain the *set* $\{a,\ b,\ c,\ d\}$

☐ whether we can arrange the members of the set in a predetermined order (we speak of an ordered set).

Just as in normal speech we need to specify the object we are talking about, we have to specify the nature of the mathematical objects we are talking about and are going to put into a set. In the language of set theory, we need to specify the universe of discourse[9] **U**, which is a rather grand general name for the objects from which we are going to build up our set. In elementary mathematics there are six different sets of mathematical objects which are frequently used as the universe of discourse **U**. They are all sets of numbers and it is worthwhile learning them and their names by heart, since they appear almost everywhere. They are given below and the list should be carefully studied, and the differences between the various definitions, names and number ranges[10] noted, since they form a basic and indispensable part of the vocabulary of every mathematician:

N   : the set of all **natural numbers** 1, 2, 3, 4, ...[11]

Z   : the set of all integers ..., -3, -2, -1, 0, $+1$, $+2$, $+3$, ...[12]

continued

---

9 The word *discourse* means *speech or language generally, conversation*. The *universe of discourse* is thus the collection of *all* the things or objects we are talking about.

10 For instance, where they start, whether the *signs* have to be included, whether the *zero* is included, the difference between an *integer* and a *non-negative number*, the difference between a *rational number* and a *real number*, etc.

11 It should be noted that the number zero *0* is sometimes regarded as a natural number.

12 This list of numbers is sometimes written as

$$0,\ \pm 1,\ \pm 2,\ \pm 3,\ ...$$

where the $\pm$ sign before the number 1 means that -1 and $+1$ are included in the list of numbers contained in **Z** (it is not read as the words 'more or less' which we use in ordinary spoken communication to indicate, for instance, that something is not accurately measured, or when we feel that a story was more or less told correctly by a friend).

---

continued

$\mathbf{Z}^+$: the set of all **non-negative integers** 0, 1, 2, 3, 4, ...

$\mathbf{Q}$: the set of all **rational numbers,** that is, all *fractions*[13] of the form p/q, where both p and q lie in $\mathbf{Z}$, and $q \neq 0$

$\mathbf{R}$: the set of all **real numbers,** that is, all possible decimals (whether terminating or not) which we sometimes think of as the set of all points on a straight line

$\mathbf{R}^2$: the set of all **points consisting of the pairs of real numbers** (x, y) **in a plane** of co-ordinate geometry.

It is worthwhile spending some time thinking about the following questions:

☐ Where do the numbers of the different universes of discourse start?

☐ For which do the *signs* have to be included?

☐ Is *zero* included?

☐ What is the difference between an *integer* and a *non-negative* integer?

☐ What is the difference between a *rational number* and a *real number*?

☐ Why are the numbers of $\mathbf{N}$ called the *natural numbers*?

☐ Why is *zero* not a natural number in the definition given above?

☐ Why are the numbers p/q with $q = 0$ specifically excluded from the numbers of $\mathbf{Q}$?

☐ May x and y be identical in the number pair (x, y)?

If these questions can be answered, then one can be relatively certain that the sets and their names have become part of one's own vocabulary. Just as the student of a foreign language must learn the required vocabulary in order to speak it fluently, the student of mathematics is required to learn the relevant mathematical vocabulary. It is also important to revise, just as the student of a language does, in order to reinforce the learning process. Many of the problems which students experience with mathematics are due to the neglect of this aspect.

These sets of numbers, or **universes of discourse U** play a central role in mathematics, and it is essential to know them by heart, since they and their symbols are used everywhere. More often than not we find that the **property** $P(x)$ which characterises the elements x of a set includes the fact that x is a member of one of the above **universes of discourse.**

It is very interesting to read about the history of numbers, since they go back to the dawn of civilisation. One of the greatest discoveries

---

13 Attention is drawn to the fact that a *fraction* refers to any *quotient* or *ratio* of *integers with non-zero denominator,* that is, by a fraction we understand not only numbers like 3/5, which are less than one, but also numbers like 2/1, 57/21 and 113/1, which are greater than one.

of mankind was that of the number **zero** (which is derived from the Arabic *cifr,* meaning *empty,* or *number*; the Arabic word is also the basis of our word *cipher,* meaning any one of the Arabic numerals). Babylonian mathematicians were the first to use the zero (meaning empty) in numbers written as the exponents of their number system (having 60 as the base[14]) to show that there was no exponent of the base in some cases. (The *number 0* should, however, not be confused with the empty set $\varnothing$.) For instance, a Babylonian mathematician wrote the following number 3; 0; 18, which is equal to $(3 \times 60^2) + (0 \times 60^1) + 18$. Their astronomers were the first to use the zero at the end of numbers, for instance

2; 11; 46; 0 = $(2 \times 60^3) + (11 \times 60^2) + (46 \times 60^1) + (0 \times 60^0)$.

All this seems to have happened before the end of the third century BC, although the notation never reached the Egyptian and Roman mathematicians at all! It is still not clear when the zero indicating the **zero of subtraction** (for instance, $10 - 10 = 0$) was invented, that is, our **number zero** which occurs in the sets $\mathbf{Z}^+$, $\mathbf{Z}$, $\mathbf{R}$ and $\mathbf{Q}$. This zero only appeared in Europe in the twelfth century via the Hindu and Arabic numeral system which was already well-established in Sanskrit from the fourth century onwards.

☐ Mathematics as a Spoken Language

When beginners look at a page of a mathematical text its complex lay-out is immediately evident – a lay-out which almost always scares them because the page looks abstract and full of strange words and symbols, and it thus seems difficult to read and to understand. Such a page usually consists of a compactly written **text** which is interdispersed with **Greek letters, symbols, formulae, definitions, signs, numbers,** and strange-looking **conventions.**[15] In addition, the layout isolates certain **equations** by placing them in a white-border so

---

14 We still have remnants of their base system in our modern world: a minute has 60 seconds and an hour has 60 minutes; our measure of angle, namely the *degree,* is measured in terms of 60 minutes, and the angular *minutes* in terms of 60 seconds.

15 The dictionary definition of **convention** is: *an agreement: established usage: fashion: in card games, a mode of play in accordance with a recognised code of signals, not determined by the principles of the game.* In mathematics the usage is rather similar to that cited above for card games – a convention is a recognised way of doing something, or writing something, etc. which everyone follows for the sake of convenience and for the sake of clarity. For instance, a function may be designated by any symbol or letter (provided that it is properly defined), but it is *conventional* in some elementary mathematics books to designate a *function* by the symbols *f, g* or *h*. It must be noted that letters like these are not exclusively reserved for functions. Nobody will ever designate the irrational number *3.14159 ....* by anything other than by the symbol π, but attention is drawn to the fact that the Greek lower case letter π is also used for other purposes.

that they stand out, as well as **tables** and **figures**. Readers are expected to know all these things to allow them to make sense of such a page. Reading such a text makes unusual demands on the reader since a multitude of things have to be remembered all at once – it almost seems as if all one's accumulated mathematical knowledge must always be instantaneously accessible in the working memory part of the brain.

The first point to realise is that any mathematical formula, equation, or expression found in an elementary course can be expanded into a full English phrase or sentence which can be read aloud – and which *should* be read aloud if problems are experienced conceptualising the material. This is especially true in the beginning, since it helps one to memorise the symbols, conventions and definitions, etc. Even the most abstract and formidable-looking equation can be read aloud because every sign, every symbol, every convention, or every equation is just a shorthand way of contracting thoughts into a compact and symbolic form. It can thus be said that mathematics can also be a *spoken* language![16]

The use of these compact forms such as symbols, equations, etc has great advantages over the rhetorical style as pointed out above:

☐ they convey **generality**
☐ they confer **clarity** to thoughts
☐ they are **easy to manipulate**
☐ they convey a **massive amount of information** in a very compact, very precise and very elegant form.

The great English mathematician Alfred North Whitehead once said the following about **mathematical notation:**

> ... *by relieving the brain of all unnecessary work, a good notation sets it free to concentrate on more advanced problems, and, in effect, increases the mental power of the [human] race.*

In fact, many a new mathematical development was catalysed by the invention of a new notation which replaced another unwieldy set of notations. A case in point here is the equality sign (=) which was invented by the Englishman, *the Physician Royal*, Robert Recorde (1510-1558), in his textbook of algebra, *The Whetstone of Witte* (1557).

---

16 With practice, a mathematician develops the ability to think abstractly in terms of symbols and condensed concepts – and does not need to fully write out a formula or equation in a full sentence, or even to speak it. If one listens to mathematicians talking, one finds that they also talk in the symbolic language of symbols and abbreviations.

He gave his reasons for inventing this sign and why he chose the particular sign as follows (and we quote his exact words, using his own typestyle conventions):

> ... Howbeit, for easie alteration of *equations,* I will propounde a few examples, bicause the extraction of their roots, maie the more aptly bee wroughte. And to avoide the tedious repetition of these woordes: is *equalle to*: I will sette as I doe often in woorke use, a paire of paralleles, or Gemowe lines of one lengthe, thus: = = = = = = , bicause not 2. thynges, can be moare equalle. And now marke these nombers. ...

This man, who has made no other contribution to mathematics which is worth mentioning, made an immeasurable contribution to the notation of mathematics by inventing the sign = , expressing equality. This step, which looks deceptively small, represents one of humanity's greatest steps towards abstract thought. The advantage of this sign was not immediately realised and it took a long time before its use became common.

As the eye becomes more and more accustomed to picking out the information contained in the symbols of mathematics, the need to read it out aloud or to write it out in a full sentence will diminish. With experience, it also becomes easier to translate what a mathematics teacher says into the much clearer shorthand of symbols, formulae and equations which are then written down in a class notebook.

It is, perhaps, necessary to give an example which illustrates how easy it is to rewrite a formula or definition into ordinary mathematical English, and thus to speak the piece of mathematics in the form of a full sentence. This can be done even by a person who does not know any mathematics – *when a list of the definitions and symbols which are used in the mathematical statement is available.*

The following dictionary list of symbols and notations must be used just like an ordinary bilingual dictionary (such as a Spanish – English dictionary) in order to translate the definition which follows below:

R	the set of **real numbers** (as defined above)
$\in$	a symbol indicating *an element of*
a	a symbol denoting a **number** from **R,** that is $a \in$ **R**
b	also a symbol denoting a **number** from **R**
a, b	a pair of real numbers, each one coming from **R**

continued

---

continued

(a, b)	an **ordered pair** of real numbers,[17] each one coming from **R**
$\{\,\}$	a pair of brackets used in the definition of a set
$\mid$	a symbol denoting the words *such that . . . holds*
P	a mathematical **property,** such as *a number smaller than the number 7*
$\{a \mid P\}$	a compound symbol indicating a **set** of numbers such that **property** P holds, where P is a specified property
f	a symbol which indicates a specific **function**
$\exists$	a symbol which indicates the words *there exists at least one* element or object
$Dom(f)$	a compound symbol indicating the words *domain of the function f*

This short list of symbols deserves a thorough study, since it illustrates how the mathematician builds up complex symbols and conventions by consecutively adding concepts and their symbols:

☐ firstly, the name of the set of numbers we are going to be working with (also called the **universe of discourse**) is specified to be **R,** that is, we are going to work with a set of numbers which is chosen from the set of all real numbers

☐ we cannot list all the numbers which occur in **R** when we talk about them, so we list them as *variables,* such as a or b and we invent a special symbol, namely, $\in$ so that we can write $a \in$ **R,** that is, we say a *is an element of the set of real numbers* **R**

☐ this is followed by specifying a pair of braces which we are going to use in further definitions, namely the **curly brackets** $\{\,\}$, but without specifying what we intend to put inside them

☐ this is followed by defining a compound symbol $\{a \mid ...\}$ indicating *the set of all elements a such that.... holds*, where the upright bar is read *such that ... holds*

☐ then we specify that we indicate a mathematical property of an element of a set by the symbol P (if there is more than one relevant property, then we can call them P_1, P_2, P_3, P_4, etc.); P may

---

17 The fact that the pair is defined as *ordered* implies that the positions where the numbers occur must be distinguishable, that is, the positions can be numbered – otherwise the elements a and b cannot be ordered. In other words (a, b) and (b, a) are considered to be different pairs, unless $a = b$.

be the property that an element belongs to the set of real numbers, then P is simply $a \in \mathbf{R}$, or if we want to say that the element a belongs to $\mathbf{R}$ and is restricted to be bigger than -1 and smaller than $+1$, then P is $a \in \mathbf{R}$ and $-1 < a < +1$.

☐ then the **property,** called P, which determines the nature of the numbers a which are selected from $\mathbf{R}$, is added to give the compound symbol $\{a \mid P\}$ which is read as *the set of all a such that property P holds*

☐ the property P must be further specified (P may be a simple property such as the fact that a **is contained in or is a member of $\mathbf{R}$** to give $\{a \mid a \in \mathbf{R}\}$, or it may be a rather complicated property)

☐ the sign $\exists$ is added inside the brackets to indicate the verb phrase *there exists at least one.*

The definition we are interested in is the following:

Definition

$$Dom(f) = \{a \mid a \in \mathbf{R} \text{ and } \exists\, b \in \mathbf{R} \text{ with } (a,b) \in f\}$$

The reader should expand this definition for himself in the form of a full sentence using the list of symbol assignments given above; it is normally easiest to write it out first, and then to read it aloud. The translation is as follows:

> The domain of the function f is the set of all real numbers a such that there exists a number b belonging to the set of real numbers $\mathbf{R}$, with the ordered pair (a,b) being a member of the function f.

Naturally, such an approach in which the reader does not really understand the relevant mathematics, may lead to errors (just as a dictionary in the hand of a person who does not understand both languages may well lead to wrong translations![18]). This is only used here as an illustration to show how the mathematician systematically builds up very complex symbols and conventions. Once the student has grasped the process by which the natural scientific language in which we think is converted into symbols, etc, as well as vice versa, mathematics becomes much easier to understand.

---

18 This is sometimes evident when one reads the text of the multilingual instructions which are supplied with digital watches or other electronic equipment!

Once the beginner gets used to the symbols and remembers their meanings, it becomes easier to read such complex sets of symbols aloud and to analyse them so that they make grammatical sense. It is sometimes found that a mathematical equation is only understood after it has been transcribed into a grammatically correct English sentence!

The beginner who starts studying mathematics – specially at university – tends to be overwhelmed by large numbers of new and even strange symbols, notations, conventions, definitions, etc. One way to survive is to set up a personal dictionary list of all such symbols, notations and definitions similar to the list given above. It is usually a good idea to use a separate notebook for this purpose. Such a list tends to become rather unsystematic when it is in the growing stage, when symbols, etc are added consecutively as they occur in the book or in the lectures; it may be worthwhile to order and to edit it after a while. This editing process has the added value of automatic revision of the theory. This list should always be kept on hand when studying mathematics – it really is indispensable! It is advisable to make such a dictionary list – even when there is one in the textbook, since a personal list contains all the necessary symbols, etc which are required by the individual.

One of the things which must be remembered is that there are only a limited number of symbols in print fonts which can be used by the mathematician. It therefore follows that some symbols are used for a variety of purposes in the various branches of pure and applied mathematics. In such cases care must be taken to display all these uses in the dictionary list; it is usually a good idea to add a simple example to make things even clearer.

The mathematician and the scientist make extensive use of foreign alphabets for their symbols and formulae – such as the Greek, the Gothic, the Russian and parts of the Hebrew alphabets. One of the most important, however, is the Greek alphabet and it is mandatory to learn this alphabet by heart so that all these symbols (both the small and the capital letters) become familiar. It is especially important to learn to pronounce the names for all these letters, since they are used in spoken communication. It is also of prime importance to learn to write these symbols quickly and correctly; practice is needed for this. The Greek upper case and lower case alphabet is reproduced below in **Table 10.1** for the convenience of the reader in the form of printed letters, together with the English equivalents of the Greek letters.

Table 10.1
The letters of the
Greek Alphabet and
their names

Letter			Name	Letter			Name
A	α		Alpha	N	ν		Nu
B	β		Beta	Ξ	ξ		Xi
Γ	γ		Gamma	O	o		Omicron
Δ	δ		Delta	Π	π		Pi
E	ϵ		Epsilon	P	ρ		Rho
Z	ζ		Zeta	Σ	σ		Sigma
H	η		Eta	T	τ		Tau
Θ	ϑ	θ	Theta	Υ	υ		Upsilon
I	ι		Iota	Φ	φ	ϕ	Phi
K	$\varkappa$		Kappa	X	χ		Chi
Λ	λ		Lambda	Ψ	ψ		Psi
M	μ		Mu	Ω	ω		Omega

□ Bracket Pairs in Mathematical Language

Mathematics makes regular use of various types of *bracket pairs* which form an indispensable part of the language and formalism of mathematics. Bracket pairs are used in mathematics in two main ways, namely:

(1) for purposes of mathematical **grammar** to determine:
  □ the **sequence** in which a complicated expression or formula is to be read
  □ which **predefined sequence of actions** is to be executed

(2) for purposes of **definition,** where particular bracket pairs are included in the definition of a concept or symbol.[19]

Both these patterns of use may appear in the same formula or expression, but they are easily distinguishable to the trained eye.

Bracket pairs are also sometimes inserted into complex formulae and equations:

(3) to make them **easier for the eye to read,** especially when a part of an equation is isolated by a bracket pair to indicate that it is

---

19 This usage was already illustrated in the definition of a set {...} and the domain *Dom(f)* of a function *f* which was added to the brief dictionary list.

going to be replaced or substituted by another variable, or even by a constant, for instance:

$$y = sin^2 \text{ A } cos \text{ B} + cos \text{ B } cos^2 \text{ A}$$

$$= cos \text{ B } (sin^2 \text{ A} + cos^2 \text{ A})$$

$$= cos \text{ B}$$

where the substitution is self-evident, since it is known that $(sin^2 \text{ A} + cos^2 \text{ A}) = 1$

(4) to **collect similar groups** of physical and other constants or variables together (especially in the more mathematical of the physical sciences like physical chemistry, physics, statistics and all branches of engineering).

The indiscriminate use of brackets may be quite confusing and care should be taken not to submit to the temptation to insert brackets almost at whim, especially when writing mathematical equations. [20] A bracket pair is only inserted in a formula if it is absolutely necessary for purposes of clarity, definition and/or grammar or to indicate the nesting pattern for instructions like multiply and add (and only then when there may be a possibility of confusion).

Single brackets (called left or right brackets) are seldom used in definitions in introductory mathematics courses. Therefore, wherever a bracket occurs in a formula or expression, the other member of the pair should immediately be searched for; the expression should not be read until all the pairs have been correctly identified.

There can be various bracket levels in an expression, that is, bracket pairs may be nested inside one another:

$$\{(1 - x^2)(1 + x) - 1\}\{[sin \ (a + b) \ cos \ (a + b) + 1] \ (1 + sin \ d)\},$$
$$\{(\ )(\ )\}\{[(\ )(\ )](\ )\}.$$

In very complicated expressions beginners sometimes find it useful to connect the members of the respective bracket pairs with thin pencil lines to accentuate what they are enclosing. One of the most important rules for the writing of mathematical formulae and expressions is that **the lines connecting nested bracket pairs may never cross.**

---

20 It is a common mistake of the beginner to insert bracket pairs almost at random in derived equations. The reasons for this are not clear, although it may be due to an unconscious desire to make the expression somewhat clearer. If the use of brackets is properly understood by the student, this common and even confusing practice will be eliminated.

Such nested brackets are always read and, hence, evaluated, starting from the centre bracket pair of each nest, using the rules of mathematical combination (about which more will be said below).[21]

As said above, brackets play an important role in **definitions** and the reader should be alert to spot differences in their usage. Various combinations of the left and the right brackets of the following bracket pairs are used for this purpose; we display most of them below, with the diamond ♦ as placeholder:

(♦)	round brackets
[♦]	square brackets
{ ♦ }	curly brackets or braces
< ♦ >	pointed brackets
<< ♦ >>	double pointed brackets

Combinations of the above left and right-hand brackets are commonly found, for instance:

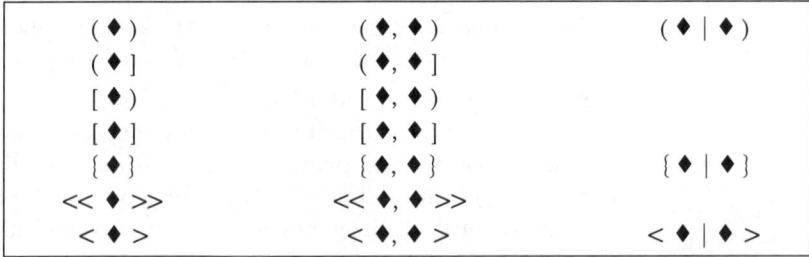

Care should be taken to enter all **definitions** which involve bracket pairs in the dictionary list and to learn them, and especially, to distinguish between them. Such definitions are usually very consistent and systematic and the reader should be on the look-out for signs indicating this underlying philosophy. As an example, consider the following two definitions, referring back to the little dictionary table of signs in the section above, while keeping a sharp eye open for the usage of the signs < (is smaller than) and ≤ (is smaller than or is

21 This mathematical expression may be read in different, but equivalent ways, for instance, keeping the brackets in for the sake of clarity, although they are not necessary in the sentence: {[The *number one* is added to the product of the *sine of the angle (a plus b)* and *the cosine of the angle (a plus b)*]; the result is multiplied by (an expression consisting of the *number one* added to the *sine of the angle d*)}; the result is multiplied by an expression consisting of the {product of (the expression which is obtained by adding the *number one* to the *negative of the square of the variable x*) and (the expression obtained by adding the *number one* to the *variable x*) from which is subtracted the *number one*}.

equal to),[22] as well as the bracket pairs ( ) and [ ] (and the implications they bring into the definitions):

An **open interval** (a, b)[23] with end points a and b is the set $\{x \mid x \in \mathbf{R} \text{ and } a < x < b\}$.

A **closed interval** $[a, b]$ with end points a and b is the set $\{x \mid x \in \mathbf{R} \text{ and } a \leq x \leq b\}$.

The point to consider is: does the variable x assume the values a and b in the case of the two definitions? The answer to this question explains the choice of names for the intervals and is useful for remembering the difference between the two definitions and the notation.

It is logical to try to extend the definitions to include the following sets which are **closed on the one side and open on the other:**

$$[a, b) = \{x \mid x \in \mathbf{R} \text{ and } a \leq x < b\}$$
$$(a, b] = \{x \mid x \in \mathbf{R} \text{ and } a < x \leq b\}$$

Hint: The usage of the signs $<$ and $\leq$, as well as the four different brackets (,), [and] in the bracket pairs are important here. The usage of the signs and the bracket pairs should be related to those in the other two definitions given above. The same question as above is pertinent here and its answer should be carefully pondered.

☐ The Verb *to be* in Mathematics, including the Signs $=$, $\equiv$, $\approx$ $<$, $>$, $\geq$, $\leq$, $\gtrless$ and their Negations

Introduction: the use of the verb *to be* in mathematical English

Some of the **mathematical symbols,** such as the $=$ sign, as well as the verb *to be,* have been used freely in this chapter without really defining them. It is, however, necessary to take a closer look at them and how they are defined and used in mathematical language. The verb *to be* and the symbols which are derived from it have very exact meanings which play a central role in mathematics.[24] This is also true for the mathematical sciences, such as applied mathematics, physics, statistics, chemistry, computer science, logic, etc.

---

22 These signs such as $<$, $>$ contain a verb as indicated; this verb must never be left out!

23 This symbol is identical with that used for the ordered pair in the list above; in actual practice, the difference in meaning usually follows from the context in which they are used; there is a small possibility of confusing them.

24 A somewhat old-fashioned analysis of aspects of the verb *to be* and the various mathematical symbols connected to it is given in this chapter – which is addressed

continued

When the symbolic language of mathematics in an equation, expression or definition is 'translated' into English, there are certain *mathematical signs* which are mapped on variations of the English verb *to be*.[25] In the written part of the mathematics text which occurs between the equations we also find that various similar phrases appear; some of them are clearly derived from the symbols, while for others there are no symbols representing or abbreviating them. We will, in the first instance, make no distinction between those expressions involving symbols and those which do not, since their grammatical functions are identical. It is interesting to look at the various ways in which this word *to be* and its variations play a role in mathematics.

It is important to realise that we do not only read our sentences from left to right, but that all mathematical statements, formulae, functions and equations are also read from left to right, with the exception of those actions which are defined by the sequence rules of the mathematical operations (+ , − , × , ÷). There are very clear conventions in mathematics about what is written on the left side of any mathematical sign involving the word *to be*, and the reader must be on the alert to spot them. One thing is clear: any argument which involves a sequence of such signs flows from left to right.

---

to the student who has neither mathematical sophistication nor proficiency in the English language. Attention is drawn to the fact that the modern mathematician and logician construct the mathematical language in terms of *arguments* (such as names, constants and variables) and *predicates* − so that the verb *to be* and other verbs become redundant. The statement *John loves Mary,* for example, may be rewritten to yield the sentence *John has the property of loving Mary,* where *John* is the argument and *has the property of loving Mary* is considered to be a *unary predicate.* Alternatively, it can be argued that there are two arguments, *John* and *Mary* and *loves,* which is a *binary predicate.* The concept of *identity* in this formalism is expressed by a special identity predicate, called = , which extracts from the many uses of the verb *is* just one − the use of *is* in the sense *is equal to.* It is felt that this approach will probably confuse the beginning student; it can, therefore, wait till the student has acquired proficiency in the use of ordinary English in mathematics.

25 The way that these symbolic signs found their way into mathematical usage, was actually just the other way around: the mathematician first used them as a type of shorthand for his thoughts − as we have indicated above for the invention of the sign = . They have, in the mean time, now acquired a life and meaning of their own. The proficient mathematician can think almost entirely in terms of these symbols and does not need to laboriously translate every symbol or every expression into its full verbal form before it can be understood. The human brain is fully adaptable and can handle this totally symbolic and almost wordless language dealing with thoughts, logic and arguments. It is an interesting experience to watch two experienced mathematicians 'talking:' they use a bewildering combination of ordinary spoken English, as well as 'spoken' symbolic mathematical language (one hears things like *Dom-ef, ef-eks, dee-ef dee-tee,* etc), and, inevitably, they move to a blackboard or a piece of paper because they have become used to thinking in terms of symbols and concepts.

There are **five main ways** in which the word **to be** appears in mathematical texts, mostly in its forms: **is** (singular) and **are** (plural). These ways are:

(1) as an **indicator of identity** between what is written on its left and what is written on its right

(2) as a **copula,** or **linking word** which links what is written on its left to what is written on its right

(3) as an indication of the **state, condition** or **property of the object**[26] **on its left** as modified by the word on its right

(4) as an **auxiliary verb**

(5) in the **passive of other verbs.**

On the surface, this usage appears to be identical to that which we find in normal or non-mathematical English. However, when one looks a bit closer at the use of the verb **to be** in mathematics, it is soon found that this apparent similarity with normal everyday usage is not actually true. Mathematics places carefully defined constraints upon each and every use of the word, whether it stands alone or in a series of carefully defined **verb phrases.** These **verb phrases** usually involve various combinations of **adjectives, adverbs, nouns** and **prepositions.**

Before we analyse the use of the verb **to be** in mathematical English, we would, however like to point out that these verb phrases cannot stand alone, and must be followed by one of the following:

☐ a **noun**
☐ an **adjective**
☐ an **adjective** which is followed by a **noun**
☐ an **adverb** which is followed by a **past participle.**

There are also **noun phrases,** like *the product of the first one hundred integers* or *the value of the sum of the integers between* − *10 and* + *10* which can either precede or follow the verb *to be.*

It is important to be on the look out for words which correspond to these types in the examples given below. In the sections which follow, we look at the five ways in which the verb **to be** functions in mathematical English, how this usage is mapped upon the respective mathematical symbols, as well as giving (and, of course, analysing) some examples.

---

26 We are not using the word **object** here in the *grammatical sense of the word* meaning the object of a sentence, but with reference to a mathematical object or name, such as *an element of a set, a variable, z,* etc.

An
indicator
of identity

There are two types of statement which involve the concept of identity, namely:

☐ the **reflexive type,** involving only one subject A
☐ the **two-subject type** involving A and B.

The **reflexive type** of statement, which states, for instance, that A multiplied by itself is equal to itself, does occur in mathematics. An example, for instance, is the following statement about an operator, called A:

A is idempotent.

The statement consists of **is + adjective.** It is similar to a statement like **Jack is stupid,** where the adjective **stupid** qualifies or describes the person who answers to the noun (name) **Jack.** In the same fashion, the adjective **idempotent** describes the operator A. We accordingly have to know the definition of something which is **idempotent,** namely:

☐ a mathematical entity is called **idempotent** when, mulitiplied by itself, it yields itself back again:

$$A \times A = A^2 = A$$

Numbers can be also be idempotent, for instance, the number one: $1 \times 1 = 1$. Another example is $a^0\ (a^0) = a^0 = 1$.

There are two different kinds of the two-subject type of sentence which states that A is identical to B, the verb phrase consisting of the combination: is + adjective + preposition. Both kinds actually mean that A and B are one and the same thing, and we need not exactly define what we mean by the phrase *are one and the same thing* at this stage, since we assume that it is always clear from the context.

The first kind of **two-subject identity statement** that we can make, refers to mathematical objects such as *sets, variables, constants,* etc. If we say that $A = B$, then we say that they are identical in all respects. The phrases 'identical with' or 'identical to' thus mean that the two objects – which we have called A and B – match exactly in all respects.

This is one of the identifying characteristics of mathematical English which makes it different from ordinary speech: all such matchings must be exact; it must be proved that all aspects of A and B match. In real life we can easily pass a quick eye over an object like a flower, and say: Oh, yes! It is identical to the one Jane has. Such a statement cannot be true in mathematics, since Jane still has her flower; some reflection shows that we are rather confused and are actually not talking about the flowers themselves, but about the species to which the flowers belong (although the person to whom we were speaking understood us perfectly!). This type of statement can never be made in mathematics, since the latter aims at being both unambiguous, as well as exact.

The second kind of **two-subject identity statement** that we can make, refers to mathematical **propositions.** The general concept of proposition – meaning a statement which will be demonstrated to be a consequence of of our axioms – is sometimes indicated by words such as: **theorem,**[27] **lemma**[28] and **corollary**[29] which are used in special circumstances.

There are relations in the **calculus of propositions** which are equivalent – meaning that they are true under exactly the same circumstances – although they are formulated in quite different ways; they are called **equivalence relations**[30] which are also indicated by the symbol $\equiv$. However, a dot is added by some authors on both sides of the sign in a statement like $p. \equiv .q$ to indicate the absence of braces. The following examples ought to be carefully studied, so that the use of the equivalence sign in the following rules of the calculus of propositions becomes clear. We use the abbreviations p, q, and r to indicate three different propositions, and the expression *not* p[31] for the negation of the proposition p in the following rules dealing with propositions, keeping in mind that *or* in mathematics always carries with it the connotation of *and*:

not (not p)	$\equiv p$
p or q	$\equiv q$ *or p*
p and q	$\equiv q$ *and p*
p or (q or r)	$\equiv (p$ *or q) or r*
p and (q and r)	$\equiv (p$ *and q) and r*
p or p	$\equiv p$
p and p	$\equiv p$
not (p or q)	$\equiv$ *not p and not q*
not (p and q)	$\equiv$ *not p or not q.*

27 **Theorem:** *a proposition to be proved;* a principle of which the truth is not self-evident, but which must be proved.

28 The dictionary (original) meaning of the word **lemma** is: a *preliminary proposition* which we intend to prove true or false. The mathematician uses the word now for a result which is proven not for its own intrinsic value or interest, but rather because its result is to be used in another, more important proof. Some lemmas are later found to be extremely important, and they acquire a name of their own, mostly derived from the name of the person who first proved it.

29 **Corollary:** an *easy inference* or a *consequence* or result.

30 They are sometimes called *logical equivalence* or *syntactic equivalence.*

31 The abbreviation for the word **not** in the negation of the proposition A, which we have written here as: *not A,* is sometimes given as $\sim A$, or $\neg A$, depending upon the text.

As an example, consider the rule $p \ and \ q \equiv q \ and \ p$, which looks quite abstract. However, if we relate it to the statement *Peter and Mary,* we can see that it is equivalent to (but not identical to!) the statement *Mary and Peter.* Let us further look at the last rule, substituting an ordinary sentence (which we all think that we understand!) instead of a mathematical proposition. Let us make the statement:

> *It is not true that she has beautiful hair and beautiful eyes.*

What we actually mean in terms of the last rule of the above set of rules, is the following: if we take p = *she has beautiful hair,* and q = *she has beautiful eyes,* then the statement in the box can be written: *not (p and q),* which is equivalent to: *(not p or not q),* using the last rule above. In other words *(it is not true that she has beautiful eyes)* or *(it is not true that she has beautiful hair).* To put it briefly, the lady has ugly hair and/or ugly eyes.

Problem 2

Rewrite the following statement into an equivalence, using the above rules:

> *It is untrue that she has beautiful hair or beautiful eyes.*

Problem 3

What is the **not** q proposition for the following proposition q:

> *Although the sun is shining, it is raining.*

Hint:

Rewrite the statement so that the word *and* occurs between two statements p and q.

The linking word

The **copula** or **linking word** in English is derived from the verb *to be:* is (singular) or are (plural). This word, which is used in the context of the sentence A is B, links two nouns A and B in such a fashion that the second noun B restricts the set of things from which A can be chosen.[32]

As an example, let us consider the pronoun (which acts in the place of a noun) **he,** which refers to something which belongs to the **male** gender in ordinary speech. The dictionary list of the nouns referring to living things, or objects which we ordinarily consider to be male,

---

32 This, again, is an old-fashioned way of introducing the subject, but this approach is used (rather than the more formal linguistic way which is common today) because this chapter is intended to be nothing more than an elementary introduction.

from which we can choose the **he** to whom we are referring, is rather large:

man, dog, horse, bear,...; bilharzia worm, ascaris worm,...; ant, fly, locust,...; avocado tree, pawpaw tree,...; whale, shark,...; hawk, pidgeon,...; reader, rider, diver,...; ... .

In the statement: *he is a man,* the linking word is forms a bridge between the indefinite word *he* with the more specific word *man;* the statement now specifies that the **he** about whom we are talking, belongs to the class of living things which we call *man* and which appears in the list of all masculine living things.[33] It is clear that we are specifying what the *he* actually is, that is, we are giving a specification which tells us what word we selected out of all the possible words referring to the male gender; we can thus call the linking word *is,* the *is* of specification.

When we use the noun *man,* each one of us has a fuzzy picture in our mind which tells us what object we consider to be a man. If we want to be more specific, we can make a list of all the attributes which make up the definition of the word *man.* Yet, we never refer to the whole *list* of attributes when we speak about a man: we are satisfied with the rather fuzzy picture which our mind's eye produces for us: we can make reasonably accurate identifications in real life on the basis of **fuzzy information.**[34]

This is not adequate for the purposes of mathematics with its stringent requirements of accuracy and unambiguousness. In mathematics, when we write: Consider the function $f(x) = 2x + 5,$ then we require that $f(x) = 2x + 5$ has *all* the attributes of the defined concept *function.* This insistent requirement of mathematics must always be kept in mind: we never ever deal with things which do not fulfil *all* the aspects of their definitions. It must be realised that a mathematical definition takes the form of a statement which includes the condition that it is true *if and only if* (which is abbreviated as *iff*) when certain stringent conditions are met.[35] For instance, when a mathematical entity does not fulfil every aspect of the definition of the concept function, then we can never ever call it a function.

---

33 It should be noted that mathematics does not deal with statements like: *He is bad* or *He is sick.* Mathematics, uses adjectives and adverbs in full sentences or in verb phrases, but they are always very precisely defined with respect to the nouns, adjectives or verbs which they modify; there are no separate mathematical symbols for them.

34 The word **fuzzy** is now part of mathematical nomenclature, since a new branch of logic, mathematics and computer science deals with the concept **fuzzy information.**

35 If a mathematician says: *I'll swim if the sun shines,* he does not exclude the possibility of going swimming when the sun does not shine. However, if he uses the stronger form of the statement, namely, *I'll swim if and only if the sun shines,* he is expressing the attitude of mind that he'll never swim when the sun is not shining.

The definition of the concept *function*[36] requires, amongst other things, that every *x-value* should yield one and only one *y-value*. Let us consider the equation for a circle $x^2 + y^2 = 1$, or $y = \pm (1 - x^2)^{\frac{1}{2}}$; if we exclude $x = 1$ and $x = -1$, then we see that every *possible* value of x which may be used, gives rise to two values of y. This means that we cannot say that we are dealing with a function in this case.

We have used the sign $=$ above without even bothering to comment, since it is so well-known: it is the symbol which is used for the linking word in mathematics.[37] If we leave out the $\pm$ sign in front of the expression y above, then we may make the statement: *the function f of x is* $(1 - x^2)^{\frac{1}{2}}$; this statement is correct, since the expression y is now single-valued over the possible values of the variable x. We can, therefore, write $f(x) = (1 - x^2)^{\frac{1}{2}}$. The expression is read as follows: *f of x is the square root of $1 - x^2$*.

At the level of linguistic sophistication which we are using, it can thus be said that the mathematical expression $(1 - x^2)^{\frac{1}{2}}$ plays the role of the second noun (which we called B) in the statement of the type $A = B$, while $f(x)$ plays the part of the first noun A. In colloquial mathematical speech when we wish to emphasise the **equality** of $f(x)$ and the expression $(1 - x^2)^{\frac{1}{2}}$, then we may say that *f of x is* **equal to** $(1 - x^2)^{\frac{1}{2}}$, which is of the form: **is + adjective + preposition** (which is discussed below). The addition of the words **equal to,** is actually unnecessary in this specific context, since the *particular* function $f(x)$ we are discussing **is defined as**: $(1 - x^2)^{\frac{1}{2}}$.

Some more examples of this type are statements like:

☐ A is an equivalence
☐ A is an injection
☐ A is a mapping
☐ the ordered pair (x, y) is a set given by $(x, y) = \{\{x\}, \{x, y\}\}$, consisting of the **singleton** $\{x\}$ *and the* **set** $\{x, y\}$; the first co-ordinate of (x, y) is x and the second co-ordinate is y.

When an author uses words like *equivalence, injection, mapping* or *singleton,* he assumes that their definitions are either known to the reader or that he is going to give a definition of the term.

We'll return later to the concept **ordered pair** which is defined above in a rather complex-looking mathematical statement. It is not necessary to understand the mathematics of this statement, but it should be analysed in terms of the sequence of *is*-statements. This is typical

---

36 The point which we want to emphasise, is the fact that the word *is* which we are discussing is actually the copula (or linking word) **is** which appears in **definitions.** Such a definition was discussed above for the case of the *domain* of a function.

37 Modern mathematicians call it *the identity predicate.*

of the concentrated style of writing of mathematics. In this rather complex-looking statement, a number of very important definitions are given one after the other. The reader who is reading such a statement is required to list them separately, to understand them, to learn them, and, above all, to remember them, because the author implicitly says that they are going to be used further on in the text – the mathematician never gives needless or useless definitions!

The indication of the state, condition or property of something

The verb *to be* is also used as an indication of the **state, condition** or **property** of the object[38] on its left as modified by the words on its right. The verb, which usually appears in its forms *is* (singular) or *are* (plural), has to be qualified by adding suitable words after it. These added words are of the types:

is + adjective
is + adjective + preposition
is + comparative adjective + conjunctive
is + adverb (which modifies adjective) + adjective + preposition
is + (indefinite/definite) article + noun + preposition
is + past participle + preposition[39]
is + adverb modifying following verb/past participle + (preposition)

is + adjective

Let us first look at the case of **is + adjective,** where the word *is* plus the added *adjective* tell us something about the concept on the left. If we, for instance, in ordinary speech say:

A is

we are saying that *A exists,* whatever *A* might mean. This type of sentence cannot be used in mathematics, where we would rather say *There exists an x such that ...* to indicate the existence of something.

If we add an adjective after the word *is,* then we are qualifying the way in which *A* exists. For instance, if we say that *A is symmetrical,* we are specifying something about the particular *A* which we are talking about and which makes it *different* from all other *A*'s. What we say makes it different from the other *A*'s is the *exactly-defined property* in mathematics which we call *symmetrical.* This means that a definition of the word *symmetrical* which applies to this particular case must

---

38 We are not using the word **object** here in the *grammatical sense of the word* meaning the object of a sentence, but with reference to a mathematical object or name, such as *an element of a set, an unknown variable, Z,* etc.

39 This form is actually the **passive form** of certain verbs like: *to contain,* followed by a preposition, for instance: *The set A **is contained in** the set B,* or in the present tense: *The set B **contains** the set A* (which has the following inverse statement: *The set A **lies in** the set B*). These verbs are also discussed separately below, since they play an essential role in mathematics.

exist somewhere in the book which we are reading, and that we must look for it if we do not remember it exactly.

The two words *is* and *symmetrical* in the sentence: *A is symmetrical,* almost function together as the verb *to be symmetrical* which describes or modifies the word *A.* We'll return to this remark when we look at what happens when we add an *adverb* to the statement.

Other examples of the qualifying **is + adjective** are as follows, where *A* in the statement *A (is + adjective)* may be a more complex statement which acts as if it were a noun:

☐ *the set* **R** *is* **infinite**
☐ *a finite subset of a finite set* **is finite**
☐ *given n* ∈ **N***, then P(n)* **is true**
☐ *a neutral element in A* **is unique** *if it exists*
☐ *the ordered pairs (x, y) and (u, v) with x, y, u, v* ∈ **R are equal** *iff*[40] *x = u and y = v.*

It is worthwhile to take a good look at the way in which these sentences are built up, as well as at the type of adjectives. The important point which must be remembered is that all these adjectives are very precisely defined somewhere in the text in which they appear: the reader should, therefore, be on the look out for such definitions; they are meant to be understood and learnt.

is +
adjective +
preposition

The **second type** is that of the combination: **is + adjective + preposition.** There are many such examples where the condition of the entity on the left side is qualified by what is on the right side. The prepositions which are used are fixed in the sense that each adjective has only one specific preposition which follows it; there are no general rules to determine what follows which, and it is best to learn them by heart, so that they can be correctly used. Examples are:

☐ *x is different* **from** *y* $x \neq y$
☐ *x is divisible* **by** *y*[41]
☐ *is equal* **to** $=$
☐ *is equivalent* **to**
☐ *is identical* **to** $\equiv$[42]

---

40 The word *iff,* which appears only in mathematics, is an abbreviation for the phrase: *if and only if.* This phrase represents a much stronger condition than the ordinary *if.*

41 The word *divisible* is not the same as *divided by.* If *x is divisible by y* it means that *x* is an *integral multiple of y,* that is, if y is divided into *x* then there is no *remainder* or *residue* (sometimes it is colloquially called *rest*). In such a case, $x|y$ would yield one of the integers from the set **Z**, for instance. *Divided by,* means that the *fraction x/y* is numerically worked out to give one element of the set of real numbers **R**.

42 This sign is also used to indicate a *congruence.*

□ *is isomorphic* **to**
□ *is maximal* **with respect to.**[43]

It is, perhaps, worthwhile to say someting more about the = sign, which means *is equal to* in mathematics. It must be noted that there are actually many different kinds of *is equal to* in mathematics! And yet we use the same sign (=) which means the same: *is equal to* for all of them. Every type of usage actually needs its own definition of what we mean by invoking the property of equality between two mathematical objects. For instance, if we say that the ordered pair of numbers (a, b) *is equal to* the ordered pair of numbers (c, d), *then* $(a, b) = (c, d)$ *iff* $a = c$ *and* $b = d$, *where we assume that we know what we mean when we say, for instance,* $a = c$, where both a and c are real numbers. If we say that the vector **a** in three-dimensional space is equal to the vector **b,** that is, $\mathbf{a} = \mathbf{b}$, then the vector **a** and the vector **b** must have the same direction and the same magnitude (whatever that may be, as defined in the theory of vectors). The equal sign = thus carries within it the concept of identity, and what we mean by the word identity being specially defined for each type of usage.

is +
comparative
adjective +
conjunctive

A comparative adjective is one that expresses comparison between two things A and B, where A and B are given in a definite order. An example is the statement:

$$x \text{ is bigger than } y^{[44]} \qquad\qquad x > y.$$

A *connective* (sometimes called a *conjunctive*) is a word which connects sentences. The conjunctive *than* is necessary here to indicate the comparison.[45] The *order* of x and y in the statement clearly shows that x (according to the writer of the sentence) is compared with y and found to be the *bigger* of the two. Another example is:

$$x \text{ is smaller than } y \qquad\qquad x < y.$$

is +
adverb +
comparative
adjective +
conjunctive

This type of phrase is more complex than any of the previous types, and we must take a careful look at the role of the adverb. Let us consider a sentence such as:

$$x \text{ is just bigger than } y.$$

The sentence actually says that the variable which we call x is bigger than the variable y, and also specifies that this difference is such that $(x - y)$ is both positive as well as not much different from zero.

---

43 The expression *with respect to* is an idiomatic expression meaning *with regard to;* it functions here as if it were a compound preposition.

44 The concept *bigger than* will further be discussed below. Here we are concerned with the general cases. It is to be noted that signs such as > include a verb part, since it means *is greater than,* and it is wrong to speak only of *greater than.*

45 Modern mathematicians tend to regard the predicate *is greater than* as an indivisible unit and they do not assign meaning to its parts.

We must remember that an adverb also can modify an adjective. It is thus clear that the adverb *just* tells us how much bigger x is than y: it is *barely bigger than y*.

The most common examples and their symbols are:

☐ *is just bigger than*	$\gtrsim$
☐ *is just smaller than*	$\lesssim$
☐ *is very much bigger than*	$>>$
☐ *is very much smaller than*	$<<$
☐ *x is almost equal to y*	$\approx$

is + article + noun + preposition

These examples and their symbols (where relevant) are actually self-explanatory:

☐ *is a function of*	
☐ *is a member of*	$\in$
☐ *is a prerequisite for*	
☐ *is the locus of.*	

is + past participle + (preposition)

As said above, this type of use of the word *is* actually refers to the passive form of certain verbs, but we treat it here because it looks so similar to the other cases above. This is only true on the surface, since the passive form of the verb which we are using here tells us that something had been done to the object A in which we are interested. The preposition which follows the past participle always specifies another noun which is involved with the verb. Let us look at an example, such as:

The set $A = \{a_1, a_2, a_3, ..., a_n\}$ is closed under multiplication.

We must remember that when we *close* something (like a box with a lid), then we take care that *nothing else* is put into it or taken out of it, since we are happy with its contents (for example, marbles). The implication of the verb *to close* in the case of our set A is that the bracket pair { } functions in exactly the same way as the lid of a box: no matter what we do with respect to multiplication with the elements of the set inside the box, nothing gets out, nor can any other element be put into the box because of the multiplications: the set is closed. The sentence thus says that when we multiply the elements of the set pairwise together, then we'll always obtain an element which is a member of the same set A.[46] The preposition *under* has the dictionary meaning: *beneath, lower; in accordance with; in the course of; referred*

---

46 This statement is not so strange as it seems, for it is true for all six the sets of numbers (called the *universes of discourse*), like N, Z and R. Take, for instance, the set A as the set N; if $a_2 = 2$, $a_3 = 3$ and $a_6 = 6$, then $a_2 \cdot a_3 = 2 \times 3 = 6 = a_6 \in N$. One can also generalise this binary product, and show that it holds for any pair of numbers in R.

to the class. The meaning of *under* in this case is just about the same as the expression *in the course of*.

There are, again, no rules to say which preposition is the correct one for a specific passive form of a verb, but each case has to be learnt. Other examples are:

- ☐ *is chosen from*
- ☐ *is contained in*
- ☐ *is deduced from*
- ☐ *is determined by*
- ☐ *is divided by* $\div$
- ☐ *is enclosed by*
- ☐ *is included in*
- ☐ *is multiplied by* $\times$

is + adverb modifying following verb/past participle + (preposition)

An adverb (such as *well, truly*, etc) modifies the past participle of the verb which follows it. The preposition can be left out of the statement, for instance, as in the complex sentence (it is actually a definition):

- ☐ *the ordered set (X, P)* **is well-ordered** *iff every non-empty subset S has a least member, where P is an ordering relation;*

or, more simply:

- ☐ *the ordered set (X, P)* **is well-ordered**

The meaning of the concept *to order well* must be defined somewhere before this statement can be made, or immediately after it is used. In this case, the definition immediately follows the statement *the ordered set (X, P)* **is well-ordered** in the form of an *iff* statement. The word **well-ordered** without the preposition thus functions similarly to that of the adjective *B* in the statement *A is B* which was discussed above, where *B* is an adjective. We can now make a further statement (which comes in the form of a conclusion) using the correct preposition, for instance:

- ☐ *the set A is well-ordered by the ordering relation P.*

Negation of these verb phrases[47]

In addition, these verb phrases are **negated** by the *adverb* **not**. The **negation symbol** is usually a skew bar (leaning to the right) through the relevant symbol; in some instances it is not customary to use a specific negation symbol, but rather to express negation in some other way by an alternative formulation, as we'll see below. For example, the negation of $=$ is denoted by $\neq$.

---

47 One can also see negation as a *connective* (just like *and* and *or*). This falls outside the scope of the present book.

Complex verb phrases

These verb phrases can, furthermore, be joined together into **complex verb phrases** using various combinations of connectives. These are:

☐ the connective **and**
☐ the connective **or**.[48]

In dictionaries the word *either* is also given as a connective, but in mathematics it never occurs alone, but as it were, rides piggyback on the connective *or*.

Extreme care has to be taken when encountering or using such cases, for the rules of logic always play a role. Examples are:

☐ *is (either) bigger than or smaller than* $\genfrac{}{}{0pt}{}{>}{<}$
☐ *is (either) equal to or smaller than* $\leq$
☐ *is (either) equal to or bigger than* $\geq$

It is clear that the presence of the connective *or* in such a complex symbol or verb phrase marks a **branch point** or **decision point** in any argument, giving rise to a **branching diagram** (more about this below). This is illustrated graphically for the case of the sign $\leq$:

$\leq$ ---------▶ OR ⟨ ◂— < — one branch of argument
                         ◂— = — other branch of argument

Whenever such signs occur, the reader must immediately translate them into **or** statements, and thus be fully aware of the implicit branching in the argument or proof. Giving attention to only the one branch will lose half of the sense of the sign and thus that of the argument (and in an examination, probably half of the marks allotted for the question too!).

If the connective *and* is added, then two signs in succession is used to express the two chained verb phrases, for instance:

☐ is bigger than ... **and** *is a member of*

 – for example: $a > 5$ **and** $a \in G$.

It is clear that the use of the word *and* in this case leads to a severe restriction of what a can be.

The verb phrases ≈ or ~

The verb phrase $\approx$ which was discussed in the previous section, carries the meaning of **inequality** in it, since it has the meaning: *is almost equal to*.[49] That is, when we write $y \approx x$, it means that the (unknown) number y (which appears on the *left*) is *almost*, but not *quite* equal to the (known) number x. In this fashion the statement $0.99999 \approx 1$

---

48 This connective *or* is also called a *disjunctive*.
49 Attention is drawn to the fact that this sign has several more roles in mathematics.

speaks for itself: the number 0.99999 is just 0.00001 smaller than the number one which is required. The statement $0.00001 \approx 0$ defines what the author means by the adverb *almost;* the small difference 0.00001 can thus be neglected. It is to be noted that the first number need not be smaller than the other; we could have had something like $1.00001 \approx 1$, where $1.00001 - 1 = 0.00001$.

The essential inequality of the numbers appearing to the left and to the right of this sign brings a certain amount of fuzziness to any such statement. In many cases the value of this **difference** $(y - x) \approx d$ does not really matter, but in many other cases it does. In such cases the author who uses the sign, will specify how big the largest difference d may be which she wants to neglect in the statement $y \approx x$. For instance, the display of some pocket calculators does not distinguish between numbers A and B which differ by an amount less than 10^{-8}; this might have implications for the accuracy of iterative or long calculations where one has to write the numbers down for use in later calculations.

The verb phrases $<$, $>$, $\geq$, $\leq$, and $\gneqq$

These verb phrases or signs, which were also discussed above, are usually used with respect to a **number system** (any one of the many) and make statements about the concepts is *bigger than* or *is smaller than* with respect to any ordered pair of numbers x and y which is in the **universe of discourse.**

It is not so easy to define the concepts *is bigger than* and *is smaller than* in normal language, except by intuition – after all we all know from experience what these phrases mean. For mathematics we would like to be a bit more precise, and it is perhaps the easiest to start off by looking at the set **Z.** These integers range from $-\infty$ through zero to $+\infty$, where the sign ∞ means *infinity.*[50] For our purposes, we can say that there are three kinds of such numbers, namely

☐ the *negative numbers* (such as .... -4, -3, -2, -1)
☐ the number *zero* or 0
☐ the *positive numbers* (such as 1, 2, 3, 4, ...).

The starting point of the chain of argument which leads to the point where one number may be deduced to be *bigger* (or, equivalently, called *larger*) than another number is the **definition** which we agree upon:

> **Any positive integer is considered to be bigger than the number zero.**

This is the fundamental point of departure: a **definition** which:
☐ stands by itself

---

50 It should be noted that *infinity* is not a number, but a word which indicates that the set of numbers is unbounded.

☐ rests on nothing (except the definition of the set **Z**)

☐ cannot be proven by any means

☐ is based upon our 'intuition' or common sense, or whatever we would like to call it.

We now need a special shorthand sign to indicate the fact that the positive integers are bigger than zero. The sign $>$ is used to indicate the verb phrase **is bigger than,** where the following generally agreed-upon **convention** is used:

> The bigger number is written on the open side of the sign $>$, and the smaller number on the acute or sharp end.

This then means, if x is any positive number which is in **Z**, that we can write:

$$x > 0.　　　　　　　　　　\text{(by definition).}$$

The statement is read: *the number x is bigger (or larger) than zero.*[51]

A second definition now follows, which helps to link any two arbitrary numbers x and y in **Z** to the number 0:

> If the numbers x and y are in **Z**, then x is considered to be bigger than y when
>
> $(x - y) > 0$
>
> that is, when the *number* $(x - y)$ is a positive number.

This is written $x > y$, and is read *the number x is bigger than the number y*. In evaluating the number $(x - y)$, we must keep the *signs* of both numbers in mind.

This definition clearly rests upon the previous definition, and it presumes the existence of the operation of **subtraction.** The procedure to arrive at the mathematical definition of the verb phrases expressing inequality is extremely interesting, since it illustrates how the mathematician arrives at definitions: they are arrived at in a logical and stepwise fashion. The concept is *bigger than* was formalised above for the set of numbers **Z;** it can also be done for the set of real numbers **R,** or, for any other set of numbers for that matter, and the reader is advised to reflect on this statement.

---

51 This statement has the correct feeling about it, since if the number of apples which belong to a boy is three and if the number of apples belonging to his friend is zero, he would certainly consider his number of apples to be larger than that of his friend. (This statement is phrased in this peculiar way to focus the attention on the *numbers* and not on the apples; the same applies to the footnote 53 referring to dollars.)

The same kind of argument can now be used to obtain the meaning of the sign $<$ which corresponds to the verb phrase: *is less than.* If *x is any negative number,*[52] then $x < 0$ (definition),[53] and if x and y are *any numbers* in **R**, then $x < y$ (read: *the number x is smaller than the number y*) if $(x - y) < 0$.

Problem 4

Are the negative numbers y *in* **R** bigger or smaller than the positive numbers x?

It must be noted that the signs $<$ and $>$ are, as it were, logical mirror images of one another. This means that the statement $x < y$ may be reversed to give the equally true expression $y > x$; if the roles of numbers x and y are permuted or exchanged, then the signs $<$ and $>$ are also interchanged.

The sign $\geq$ means *is bigger than or is equal to,* while the sign $\leq$ means *is smaller than or equal to.* For instance, the expression $x \geq y$ means for the purposes of the mathematical argument that it does not matter whether the number x is *either* bigger than the number y *or* equal to the number y; it must be noted that this statement indirectly says that the number x may *never* be taken to be smaller than the number y.

The sign $\gtrless$ means *is bigger than or is smaller than;* again it carries the connotation that in the statement $x \gtrless y$ there is no possibility that the number x may ever be equal to the number y.

As said above, the introduction of the signs $\geq$, $\leq$, and $\gtrless$ containing the logical *or* statement, leads to a branching in any chain of argument.

Problem 5

Reflect on the reason why the statement $x \neq y$ for $x, y \in$ **R** can *never* be replaced by one of the statements $x > y$ or $x < y$. Can the sign $\neq$ replace the sign $\gtrless$ in the statement $x \gtrless y$?

Hint: Read the definitions of these signs very carefully!

---

52 This statement should be read and considered very carefully: it says that x is *any negative* number, that is, the number which it represents carries a *negative sign with it,* for instance -5. This means that $-x = z$, where z is the corresponding *positive* number; in the case of the number $x = -5$, it follows that $-x = -(-5) = 5 = z$. The numbers -5 and 5 are quite different numbers.

53 Again this statement has the correct feeling about it: if a person owes somebody a number of dollars corresponding to the number three, she certainly has a smaller number of dollars than she would have had if she had only owed the number corresponding to zero dollars.

Problem 6
Why are there no negations of the verbs discussed in this sub-section?

☐ The Usage of some other Verbs in Mathematics

Introduc-
tion

Mathematics makes use of a wide variety of verbs in their normal dictionary meanings and it is not necessary to comment upon those. However, there are a number of normal-looking verbs which are used in a way that is peculiar to mathematics and we take a look at their meanings below. There are, furthermore, those verbs which are only used in mathematics; their use is usually very clearly explained in mathematical texts so that we need not give attention to them here.

The verbs we have chosen to discuss in this section all occur with reasonable frequency in elementary mathematical texts, but their usage is rarely explained. The failure to grasp their meanings may lead to real mathematical problems, especially for persons whose mother-tongue is not English. We have found a dictionary such as *Chambers's Twentieth Century Dictionary* (W & R Chambers Ltd, London, 1952; later editions are available) very valuable for the verbs below; the reader is urged to use a similar dictionary to look up unknown words.

The verb
to let

The verb *to let* usually means *to allow, to permit, to grant (something to somebody)*. It is also often used:

☐ as an auxiliary verb with *optative effect* (that is, expressing a wish or desire) with the meaning *to behave so as to give an impression of something; to make something appear (as if)*
☐ as a verb with *imperative effect* (that is, giving an instruction or order).

Mathematics makes extensive use of this verb in statements such as:

Let *us divide first by b, to obtain the following result* ...

☐ the verb *let* has here the meaning of **should,** and it simply indicates that before anything else is done to the expression, that the author wants us to divide by *b* first.

Let *us recall the definition of an integral before we proceed* ...

☐ the verb *let*, together with *recall* indicate that it is *imperative* that we **know** the definition of the concept *integral,* and points out that we really should understand it before we proceed any further; the author is thus actually saying that we should **revise** the concept of *integral* – he is giving us a tip that he is going to make use of the *definition* of an integral in what follows.

Let *f be a function* ...

☐ the author is saying that she is now, for the purposes of the ensuing argument, going to **assume** *f* to be a function having all the defined properties of a function; the argument which follows such a statement usually takes the form of a proof (see below) in which is proven that the assumption is either *true* or *false*.

Let *f(x) = 2x.* ...

☐ the author is saying that she now considers a specific function which maps *x* to 2*x*, that is, she is going from the **general** to the **specific** by picking an example.

If *we let δx go to zero, then* ....

☐ the author is telling us that the variable *δx* can be varied by us (in our mind's eye) in such a fashion that we imagine it to become closer and closer to zero, abbreviated by *δx → 0*; there is even a special notation for this, which we call the **limit** (sometimes abbreviated as **Lim**):

$$\text{Lim}_{\delta x \to 0} \ f(x + \delta x)$$

where the notation $f(x + \delta x)$ means that we take $f(x)$ and wherever *x* occurs we replace it with $(x + \delta x)$; we then calculate the limit of the resulting function when $\delta x \to 0$.

Provided we **let** *x = a′ then* ....

☐ the author is telling us that *x* is taken to be equal to *a′* for the purposes of the argument which follows (it is usually explained just *why* the author makes this assignment).

It is thus clear that this little verb plays an important role in mathematics and that the reader should be fully aware of the range of meanings which are attached to it by the mathematician. The word is never used in a loose context, but always identifies one of the salient features of the mathematical argument.

The verb
to assume

This verb is often used in mathematics and it has more or less the same meaning as when used in ordinary speech:

☐ *assume* (transitive verb) to adopt (something, a point of view, a convention), to take (something) for granted, to pretend to possess (certain properties, something), to pretend that (something) is something else.

The various uses in mathematics are, for instance, to be found in the following sentences:

Without loss of generality we may **assume** *that b is a real number* ...

☐ the reader is actually **given the instruction** in this statement (using the subjunctive mood) that *b* is taken to be a real number, most probably, because it is 'easier' to work with; additional information is also given to the reader, namely the fact that it does not really matter for the purposes of the argument whether b is a real number or not.

*The theorem above does not **assume** anything about the existence or non-existence of a derivative at the point d itself ...*

□ the attention of the reader is drawn to the fact that some property was not used in the derivation of the theorem – in this case, the fact that a derivative exists or does not exist at the point d did not play a role in the proof of the theorem: it is totally irrelevant for the derivation.

The verb to consider

The dictionary definitions of the verb *to consider* are as follows:

□ *to consider* (transitive verb[54]) to look at (something) attentively or carefully, to think about (something) or deliberate about (something)

□ *to consider* (intransitive verb) to think seriously and carefully, to deliberate.

The use of the verb *to consider* in mathematics is almost restricted to the transitive, that is, it always refers to a grammatical object, for instance:

*We **consider** the function $y = |x|$, where $|x|$ is called the **absolute value** or **modulus** of the number x ...*

□ the reader is **informed** that we are now going to investigate the function $y = |x|$; at the same time, the definition $|x|$ is also given.

*We first **consider** positive values of x ...*

□ we are informed (the phrasing of the statement is such that the author tells us about it without actually having to write it down!) that there are several values of x which we can consider, and that the first values we are going to turn our attention to, are the positive values.

The verb to show

This verb is used in mathematics in its dictionary meanings:

□ *to show* (transitive verb) to present to view, to prove, to indicate, to cause to be seen or known.

Its sense is mostly identical with that of the verb *to prove*. This means that the same strict rules of proof also hold for *show*; in some cases the verb has the effective meaning of *to calculate, using the recipe given*;

---

54 A transitive verb is a verb which has an object; an intransitive verb does not have an object.

in some cases such a calculation may constitute a proof. Some examples of its usage are:

> **Show** *that the following functions are one-to-one ...*

☐ the functions which are given should be taken one at a time and the one-by-one test conditions (which are known from the text) applied to each one in order to determine whether it fulfils the defining conditions or not.

> *The function now becomes* $y = -3 (x - \frac{1}{2})^2 + \frac{7}{4}$. *This* **shows** *that its vertex is* $(\frac{1}{2}, \frac{7}{4})$ ...

☐ this statement means: when the function is compared to the conditions laid down by the definition of the concept *vertex*, that the conclusion can be drawn that the vertex of the function occurs at the numerical values given.

In older texts – mostly from the Victorian era or from the beginning of this century – an archaic variant of *to show* is used, namely, *to shew*, with the imperative *shew* and past participle *shewn*.

The verb *to prove*

The verb *to prove* and the noun *proof* are often used in everyday speech and we 'prove' many things to our friends by delivering the 'proofs'. For instance, we prove the fact that Jack goes out with Jane by showing a photograph of them taken under the Elm tree in front of the City Hall. Or we prove the fact that Cape Town is further away from Pretoria than Durban is by adding all the relevant road distances on the road map. We prove the fact that it was James who took Mary's pencil and not Peter by inspecting their pockets. Or the fact that Jim has committed a crime by confronting him in court with the relevant evidence, as well as with the laws which he broke. These everyday examples show that the verb *to prove* involves a statement about something which is then verified to be either true or false by an *investigation* which supplies the *evidence*. In some cases, the evidence (for example, the photograph of Jack and Jane) is enough to convince us. In other cases, for instance, in the case against Jim, it might be necessary for the prosecution to build up the case in a series of steps, each step being accompanied by its own proof or evidence; the logical conclusion drawn from the sequence of steps is then that Jim is indeed guilty of breaking a law. The verb *to prove* is used in mathematics in the same general way: a statement (for instance, a theorem or a lemma) is made which has then to be verified (or not, as the case may be) by an investigation which displays the 'evidence' in a stepwise fashion. The only evidence, however, which is accepted in a mathematical proof is facts which are known to be true from other proofs which have been previously proven to be true. Each step in a mathematical proof is accompanied by such 'evidence'. It thus follows that any examination question which starts with the imperative 'prove that . . .' must be answered in a sequence of steps, each step being accompanied by its mathematical 'evidence'. Care must also be taken that the proof is 'complete' (that is, no essential step is left out).

The verb to find

The use of the word is that of the familiar verb *to find* which we all know, but superimposed on its range of meanings is, again, the requirement of *proof*. An example is:

> **Find** *the inverse of the function* $z = 3x - 2$

□ the instruction is actually to **calculate** the inverse of the function given, using the procedure which is described in the text, showing each step clearly, and giving the reasons for each step; this type of instruction usually calls for a given procedure to be applied step by step, assuming that the person executing it understands the background theory (in this case, what the term *inverse function* means) and can apply it systematically to a problem.

The verb to describe

This verb *to describe* has some of the meanings of the verb *to show*, although we must remember that the given description must be mathematically correct, showing all the steps and the reasons for taking them. An example is:

> **Describe** *how you would obtain the graph of* $y = f(|x|)$ *from that of* $y = f(x)$

□ the imperative *describe* here means that one must show all steps in the calculation to convert the one into the other, giving the reasons for each step.

The verb to investigate

The verb *to investigate* in ordinary English means *to search for or to inquire into with accuracy and care*. This means that the investigating procedure or rules are not specified, leading to a kind of fuzziness of the meaning. In mathematics, however, the imperative '*investigate* something' is irrevocably tied up with the mathematical rules of proof and operation which occur in that particular branch of mathematics. After the investigation a conclusion or report is expected, and this aspect is included in the meaning of the verb.

> *We now* **investigate** *the equation* $3x^2 - 3x - 1 = b$ *for different values of the constant* b ...

□ the instruction is to determine how the properties of the expression depend upon different possible values of the constant b and to determine what happens (say, to the solutions, or to the graph), using the given rules and to draw and to formulate some (general) conclusions.

The verb to suppose

The dictionary meaning of the verb *to suppose* is: to incline to believe, to guess, to imagine, to assume provisionally for the sake of an argument or proof which is then proved to be true or false, to expect something in terms of a set of rules or axioms. The range of meanings which we find in mathematics agrees with this. An example of its usage is:

> **Suppose** *that the function is defined on an interval I* ...

☐ the author is telling us that she assumes that the function *f* is defined on an interval called *I* and she is going to make use of the fact in the subsequent argument; for all intents and purposes, she is saying that *f* is actually a function in the interval *I* – until she proves it to be either true or false.

The verb *to determine*

The dictionary meaning of the transitive verb *to determine* is:

☐ *determine* (transitive verb) to put terms or limits to (something), to fix (something), to settle (a question), to define (something), to decide (something).

Its mathematical usage falls within these bounds, with the added rider that the meaning includes the concept *rigorous*. This means that when the mathematician says that one must determine something, he actually means that one has either to prove it using rigorous mathematics, or one has to do some sort of calculation, using a prescribed procedure. An example of such usage is:

> **Determine** *the turning points or extrema of the function f* ...

☐ the instruction is to calculate the turning points in a step by step procedure (which is prescribed by the theory) and to report them in a clear fashion.

The verb to denote

The dictionary meaning of the transitive verb *to denote* is:

☐ *denote* (transitive verb) to indicate by some sort of sign, to signify or to mean (something), to indicate objects (which are collected in a class).

The mathematical uses of this verb fall within the dictionary meaning of the word. Very often the imperative form of the verb is used when an author wants the reader to use a certain symbol or terminology, for instance:

> **Denote** *the number of people in England by L.*

Very often, this verb is combined with *let*, for instance:

> *Let L* **denote** *the number of people living in England.*

The verb to find

The verb *to find* is often used in mathematics; its dictionary meaning is:

☐ *find* (transitive verb) to come upon or to meet with (something or someone), to discover (something) or to arrive at (some conclusion or some place), to succeed in getting (something).

In mathematics it usually appears in an imperative sentence such as:

Find *the solution of the following equation* ....

☐ the instruction is actually *to solve the equation by the procedures given in the text which are based upon the underlying theory which was discussed;* the instruction usually includes the point that the 'finding' must be done in an orderly and stepwise manner, each step being substantiated by referring to a particular point of the theory.

The verb to sum

In normal school English we do our 'sums', that is, we do the problems set by the teacher which illustrate the relevant theory which was done in class. This meaning has just about disappeared in mathematics, although we do speak of the *sum* of two numbers when we mean that they are *added together*. The mathematical operation which is indicated by the symbol or sign ' + ' is that of the operation of **addition** which we have met several times in this book and which features again below.

There is, however, one special use of the verb *to sum* which is common in the theory of series. **A row of numbers** (also called a **sequence**) such as:

$$1, 5, 2, 8, 4, 3, 7, 6, 9$$

is turned into an **addition sum** by inserting an addition sign between each pair of numbers:

$$1 + 5 + 2 + 8 + 4 + 3 + 7 + 6 + 9 = 45$$

However, if there is a **numerical relationship** between **each consecutive pair of numbers** in the row of numbers to be added (it is almost as though the first two numbers determine the third number, the second and the third the fourth number, and so on – we have a row of numbers which is ordered by a numerical rule) then the insertion of an addition sign between each such pair, turns the addition sum into a **series**. A series may have any number of terms, but there are two main types, namely, a series with an unlimited number of terms (we call it an **infinite series**) and a series with a limited number of terms (called a **finite series**). Examples are, respectively, the infinite series consisting of the sum of all the numbers in the infinite set of natural numbers **N** and the series consisting of the sum of the first nine numbers of the set **N** of natural numbers:

$$1 + 2 + 3 + 4 + 5 + 6 + 7 + 8 + 9 + \ ....$$
$$1 + 2 + 3 + 4 + 5 + 6 + 7 + 8 + 9.$$

There is a *constant difference of one* between each pair of numbers in both series, that is, the first and second number, as it were, determine the third, and so on.

It is natural to enquire what the **sum** of a **series** is, that is, if we add all of the terms together, how 'much' is it? Let us look at the second series, which will give us a **partial sum** of the first series, since we only sum it over the first nine terms. It is important to realise that the sum of such a series is just one simple number, no matter how complicated the expression may look.[55] It is not so cumbersome to write out every **term** of this rather simple series, but it does become very tedious to do so when the terms become complicated expressions themselves (and it becomes very difficult to get an overview of the resulting series). Mathematicians have, therefore, devised a *nomenclature* and a *symbol* to help them to contract this description, as well as to make a series easier to comprehend in its entirety. If we look at the series above, we see that it has:

- ☐ a **beginning member** (the number one in this case)
- ☐ a **last member** (the number nine in this case)
- ☐ nine **members** or **terms**
- ☐ eight **addition** signs between the nine terms
- ☐ a **difference** of one between the consecutive terms
- ☐ an **answer** (called the **sum** of the series).

In order to contract the whole little series into just one term, we need to make use of the information of the first five items above. We do this by inventing a special sign that says that we *add all the terms in the series together*. This sign is conventionally chosen to be the Greek capital letter *sigma* Σ (which is the first letter of the word *sum*). This sign contracts all eight addition signs into its symbol. But we must have something such as a general symbol, upon which this *sigma* sign can work. If we call this variable $n \in \mathbf{N}$, then we can write $\Sigma\, n$ for the above series. But we have to tell the *sigma* sign just which elements of **N** we are interested in to sum. We do this by adding underneath the sign (or just to its right as a subscript) where we would like to start with the row of numbers, and at the top of the sign where we would like to stop. The sign then looks something like this:

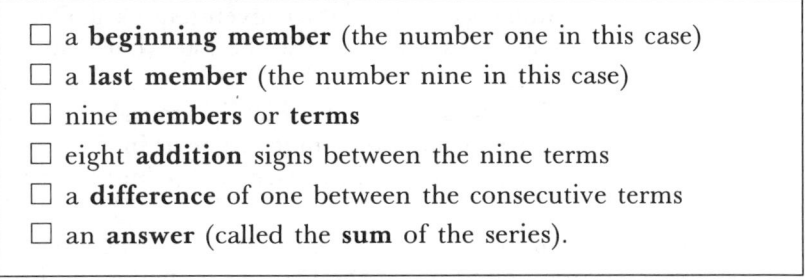

$$\sum_{n=1}^{n=9} n$$

---

55 There is a story that the teacher of Gauss, one of the greatest mathematicians of all times, gave his primary school class the sum to calculate the sum of all the integers from 1 to 100, thinking by himself: That will certainly keep them busy! Gauss gave the answer 5050 within seconds, because he realized that $1 + 99 = 100$, $2 + 98 = 100$, ..., $49 + 51 = 100$, so that the answer is reduced to $49 \times 100 + 100 + 50$.

and we **agree** that it **represents** the little series above:

$$\sum_{n=1}^{n=9} n = 1 + 2 + 3 + 4 + 5 + 6 + 7 + 8 + 9.$$

The sign thus says that we start the series at $n = 1$; we then **increase the value of n by one** for the next term and add it to the first, and so on, till we reach the last term with $n = 9$ and add it to the sum of all the others. The summation convention thus allows us to contract the whole series into just one term; conversely, if the general formula is available, the conventions used allow us to regenerate the whole series. We also use the following **convention that**:

> the *running index* (which is a variable index which allows us to step from one term to the next) in all these summation signs increases by one for each succeeding term in the series.[56]

In more complicated series, there may be more than one summation sign nested under the first, each one having its own running index which steps by one for each term; in such cases extreme care has to be taken to identify the nesting pattern and the ranges of the running indexes.

The running indices usually come from the set of natural numbers **N,** or from the set of integers **Z.** Some of the series are **infinite,** that is, the terms continue to be added without ever stopping.

There are two main types of **infinite series,** namely:

☐ a **converging series,** where the sum of the series approaches a constant value in the limit as the running index approaches infinity

    ☐ the **rate of convergence** of such a series is of importance, since some converge very slowly (we give an illustration below), while others converge faster (which is the ideal, since it is less trouble to sum them)

☐ a **diverging series,** where the successive terms cause the sum not to converge to a constant value, but either becomes very big, or fluctuates.

In some cases it may be feasible to **truncate** the series (we speak of a **truncated series** in such a case). The word *truncate* is derived from the Latin word *truncus,* meaning *maimed;* the meaning in mathematics is *to cut short.* This means that a decision is made that the rest of the

---

56 In some cases, one can also *step backwards*; this is often used in computer applications. This is indicated by exchanging the upper and lower limit indices.

terms after the truncation point can be neglected, or that they are not required. The finite series which we discussed above, is a truncated series of the infinite series Σn.

Mathematicians have developed some very elegant ways of determining the sums of **converging infinite series,** but we cannot go very deeply into further details here. Although, as a further illustration, let us look at an **infinite series** which converges to the numerical value *log 2:*

$$\frac{1}{2} \sum_{x = 1}^{x = \infty} [(-1)^{x+1}/x] = \frac{1}{2} [1 - \frac{1}{2} + \frac{1}{3} - \frac{1}{4} + \frac{1}{5} - ...]$$

$$= 0.3010299... = log \ 2$$

The values of the successive terms in square brackets $[(-1)^{x+1}/x]$ and their partial sums are:

x	$[(-1)^{x+1}/x]$	partial sum
1	1	1
2	− 0.500 000	0.500 000
3	+ 0.333 333	0.833 333
4	− 0.250 000	0.583 333
5	+ 0.200 000	0.783 333
6	− 0.166 667	0.616 666
7	+ 0.142 857	0.759 523
8	− 0.125 000	0.634 523
9	+ 0.111 111	0.745 634
10	− 0.100 000	0.645 634
11	+ 0.090 909	0.736 543
12	− 0.083 333	0.653 210
13	+ 0.076 923	0.730 133
14	− 0.071 428	0.658 705

It is clear that this series converges *extremely slowly*: after 1000 terms the value of the added term is 0.001; after one million terms the value of the partial sum is 0.602058, whereas the value of the sum is just half of this, that is, 0.302029.

The sign Σ thus represents a whole sequence of operations, and in this case we have a converging infinite series. Let us reflect a bit on the way in which these conventions contract a whole verbal sequence into one abbreviated form, using the general expression for the series, the sum of which represents *log 2*. Attention is especially drawn to the use of the factor:

$$(-1)^{x+1}$$

to determine the sign of any term. The values of the running index x which make $(x + 1)$ odd make $(-1)^{x+1} = -1$ that is, the xth term becomes negative. Conversely, if $(x + 1)$ is even, then $(-1)^{x+1} = +1$, so that the corresponding term is positive. This mathematical trick is frequently used to determine the signs of terms in general equations. Let us now go stepwise through the calculation procedure.

We initiate the calculation procedure by starting with $x = 1$ and go through the following steps which are defined by the formula $[(-1)^{x+1}/x]$ and by the Σ-sign:

☐ initiate with $x = 1$ to find the value of the first term

☐ substitute 1 for x in the formula $[(-1)^{x+1}/x]$ and calculate:

* $(-1)^{x+1} = (-1)^{1+1} = (-1)^2 = 1$
* $(1/x) = (1/1) = 1$ and thus:

  * $[(-1)^{x+1}/x] = 1\cdot 1 = 1$

☐ place the last number which represents the first term of the series $[(-1)^{x+1}/x]$ in an answer block

☐ increase the value of x by one to two to find the x-value of the second term, that is $x = 2$

☐ substitute 2 for x in the formula $[(-1)^{x+1}/x]$ and calculate:

* $(-1)^{x+1} = (-1)^{2+1} = (-1)^3 = -1$
* $(1/x) = (1/2) = 1/2$ and thus:

  * $[(-1)^{x+1}/x] = (-1)\cdot 1/2 = -0.5$

☐ take this answer and add it to the number in the answer block and replace the new number in the answer block which now represents the sum of the first two terms of the series $\Sigma\ [(-1)^{x+1}/x]$.

It is clear that we are going to go through exactly the same steps for the next terms, but we cannot ever come to the last term; for practical purposes, however, we can stop for example after about ten billion terms. Then we'll have the sum of the series in the answer block, namely, $0.3010299\cdots$. The cyclic calculation process which we have described, is called iteration, which comes from the Latin word *iterare*: to repeat, to do a second time. We say that we have gone through many **cycles** or **loops** of the iteration which was initiated by $x = 1$ and ended by $x = 10^{10}$.

We can convert the process which we have described in words into an algorithm. The word algorithm, which appeared after 1957 in scientific nomenclature[57] dealing with computers, refers to a formalised

---

57 The similar word *algorism*, meaning *doing arithmetic with Arabic numerals* was already known in the Middle Ages, and the word *algorithm* is derived from it and became a very specialised term in computer science and in mathematics. The word *algorism* is a corruption of the last name of a famous medieval Arabic mathematician, *Abu Ja'far Mohammed ibn Mûsâ al-Khowârizmi*, who lived around AD 825 and who wrote a book *Kitab al-jabr w'almuqabala* (Rules for restoration and reduction). Incidentally, our word *algebra* is derived from a corruption of the last word of the Arabic title of his book! More about *algorithms* can be found in the book by K Knuth, *The art of computer programming, volume 1, fundamental algorithms*, Addison-Wesley Publishing Company, Reading, Massachusetts, 1972.

calculation procedure which is written out step by step. Such an algorithm is sometimes (especially in older textbooks) accompanied by a flow chart, or graphical way of summarising the consecutive steps. We can contract the above calculation to give the sum of this series by the following algorithm, which includes a cycle or loop in the calculation:

STEP 1	begin
STEP 2	make certain that A = 0
STEP 3	make certain that B = 0
STEP 4	make certain that x = 0
STEP 5	make certain that x = 1
STEP 6	calculate $A = [(-1)^{x+1}/x]$
STEP 7	calculate (B + A) and call the sum B again
STEP 8	increase x by 1 by setting $x = x + 1$
STEP 9	if $x = 10$ go to step 10; if $x < 10$ go to step 6
STEP 10	$C = \frac{1}{2} * B$
STEP 11	print C
STEP 11	stop

The entity called *B* is just the 'calculation box' where we accumulate the value of the running total, and *A* is the value of each succeeding term of the series of *log 2,* while C is the final answer. We accumulate the values of the successive *partial sums* in box *B*. The cycle or loop starts at STEP 6 and ends at STEP 9.

It is interesting to work through this simple example of an algorithm to become familiar with the concept, and to develop a feeling for the process which takes place in the mathematical term: **summing a series.** STEPS 2 to 4 are necessary to ensure that all values of the variables are zero at the beginning of the calculation (a horrible thought: what would happen to the answer if any variable such as *A* or *B* is not zero at the beginning!) STEP 9 is an essential **exit step** to ensure that we do not go into what is called an **infinite loop** where we just continue on and on without ever ending. As soon as *x* exceeds the upper or end value of the variable *x* (we picked only *eight cycles* in this case, although we should actually continue till the series has converged to a stable answer), the calculation stops.

☐ Answers

Problem 1

The starting point of the proof is that the author assumes that the reader knows that:

☐ the number $(3 + {}^3/_4)$ is equal to the number $(4 - {}^1/_4)$ (i)

☐ their squares are also identical   (ii)

☐ $(+4)(-{}^1/_4) = (-1)$ (iii)

☐ the following identity holds for all real numbers a and b
    $(a + b)(a + b) = aa + 2ab + bb.$ (iv)

Step 1

Expressing (i) and (ii) in the form of an identity, it follows that

$(3 + {}^3/_4)\,(3 + {}^3/_4) = (4 - {}^1/_4)\,(4 - {}^1/_4).$

The next step is to calculate the two sides of the identity separately.

Step 2, Left-hand side

Using the identity (iv) results in

$(3 + {}^3/_4)\,(3 + {}^3/_4) \quad = 9 + 2({}^9/_4) + {}^9/_{16}$
$= 14 + ({}^1/_{16})$

Step 3, Right-hand side

Attempting to use the identity (iv) together with (iii), it is found that

$(4 - {}^{1}/_4)\,(4 - {}^1/_4) = 16 - 2 + (-{}^1/_4)\,(-{}^1/_4)$
$= 14 + (-{}^1/_4)\,(-{}^1/_4).$

Step 4

From the assumptions it now follows that

$14 + {}^1/_{16} = 14 + (-{}^1/_4)\,({}^1/_4).$

This means that

$(-{}^1/_4)\,(-{}^1/_4) = ({}^1/_4)\,({}^1/_4) = ({}^1/_{16}).$

Conclusion

☐ Multiplying a number $(-{}^1/_4)$ by the number $(-{}^1/_4)$ gives the *positive* number $({}^1/_{16}).$

Notes

(1) The proof is not general, and it cannot be said that the product of any two negative numbers is a positive number. It can, however, be speculated or

continued

continued

conjectured[58] that this will be true because the same procedure can conceivably be used for all numbers. The author of this fourteenth century textbook of arithmetic assumes that his proof holds for all negative numbers since it holds for one specific case.

(2) This is an example of a mistake which beginners often make, namely, to give a proof for a specific case when a completely general proof which would cover all possible real negative numbers was asked for; this aspect will be taken up in the section dealing with the verb to prove.

Problem 2

The second-last rule of our propositions apply here, and our equivalence is:

It is untrue that she has beautiful hair and it is untrue that she has beautiful eyes.

Problem 3

The statement as it is written, expresses a contrast to what is expected. We can usually rewrite such sentences in such a way that the word *and* occurs between the two parts. In this case, it means that the sentence we want to negate is:

*the sun is shining **and** it is raining.*

The negation is thus of the type we called **not** (*p and q*):

the sun is not shining or it is not raining.

Problem 4

It is always a good idea to subtract the one number from the other and to calculate $(x - y)$, keeping the rules of multiplication of signs $(-)(-) = (+)$ and $(-)(+) = (-)$ in mind. The next step is to determine whether the result is

☐ negative (smaller than zero)
☐ zero
☐ positive (larger than zero).

The next step is to draw a conclusion about whether $x > y$ or $x < y$ according to the definitions given above.

continued

58 The verbs *speculate* and *conjecture* are often used in mathematics to indicate that the statement which the author is now going to make has not yet been proved. They are synonyms in mathematics and carry the meaning *to make an intelligent guess*. The meaning of these verbs is thus just the opposite to that of the often-used verb *to infer* which indicates that some conclusion was drawn from a mathematical proof.

continued

In order to do this, the question states that y is a negative number (that is, y can be any one of the numbers ..., -4, -3, -2, -1) and that x a **positive number** (that is, it can be any one of the numbers *1, 2, 3, 4, ...*) This implies that $-y$ is a positive number z, that is

$$x - y = x + z > 0,$$

since the sum of two positive numbers is by definition always larger than zero. Comparing with the definitions, it is seen that the positive number x is larger than the negative number y, or $x > y$. This also follows from the last inequality by adding y to both sides of the inequality:

$$x - y + y > 0 + y$$

$$x > y.$$

Hint: A specific case such as $x = 5$ and $y = -4$, is usually convincing, but it must be remembered that it is not a general proof.

Problem 5

The statement $x > y$, for instance, specifically says that the number x must be bigger than the number y, while the statement $x \neq y$ simply says that these numbers are different, that is x may be larger or smaller than y (it, therefore, inserts an *or* in the argument where there should be none). If $x \neq y$ we do not know whether $x < y$ or $x > y$. The same holds with respect to the statement $x < y$.

For the same reason $x \neq y$ cannot be replaced by $x > y$.

Yes for $x, y \in \mathbf{R}$ we can replace $x \geq y$ by $x \neq y$ since both furnish us with the same information.

Problem 6

Four of these signs come in pairs, namely $\{>, <\}$ and $\{\geq, \leq\}$. Clearly $>$ and $\leq$ are negations of one another, and $<$ and $\geq$ are negations of one another for $x, y \in \mathbf{R}$. If the roles of x and y are interchanged, the respective signs $\{>, <\}$ or $\{\geq, \leq\}$ of each pair are also interchanged. The negation of $x \geq y$ is $x = y$. There are, however, some specific cases where authors use the negations of the signs $>$ and $<$, by putting a vertical bar through the sign.

□ Concluding Remarks

The mathematician is perhaps the most sensitive of all scientists to the use of language, and we tried to emphasise this in this chapter. Every word 'counts' for the mathematician, and every mathematical term, sign or abbreviation which is used, has a precise meaning. It is impossible to do justice to all these aspects in a brief chapter and we have tried to give attention to some concepts which we think may be of use to a student whose first language is not English.

We explained the use of some mathematical signs and symbols as well as their relationship with written and spoken English, but we did not neglect to point out that alternative interpretations exist. The reader is urged to pay special attention to sections in textbooks dealing with these aspects.

11

An Introduction to the Language and Terminology of Chemistry

☐ Introduction

Chemistry is one of the oldest of the natural sciences. It originated thousands of years ago from our desire to manipulate our physical environment to our own advantage. Chemistry is still today helping mankind to adapt to the environment and it is not possible to contemplate modern society without acknowledging the role played by chemistry. With the passing of the centuries we also developed a desire to understand the physical world and the way in which it is built up; we also would like to know how and why it changes with time and with changing conditions.

The language of modern chemistry reflects some of this long history because chemists still use some of these 'old' or archaic terms to communicate with other chemists, especially in the laboratory and in the lecture room. Old words such as *acid, salt, alcohol, water, aqueous solution, arsenic,* etc. are widely used and every chemist understands them without any trouble. However, other old words such as *muriatic acid, washing soda, blue vitriol, Turnbull's blue,* etc. have disappeared from the chemist's scientific vocabulary. Why has this happened?

One of the main reasons for the disappearance of these words is the fact that chemistry is now a *science* which requires that both its fundamental concepts and definitions, as well as its vocabulary and nomenclature (*names* of compounds and processes, for instance) be *precisely defined.* The language of a science must be such that no *linguistic misunderstandings* between chemists can occur.

The chemist also uses words and phrases of 'ordinary' literal and even colloquial English, but it should be realised that new meanings are sometimes attached to familiar words and phrases. For instance, the adjective 'volatile' modifying the name of a person tells us that the person is very easily excited. The same adjective, when applied to the compound called ether (which is a liquid at room temperature), tells the chemist that the liquid quickly disappears and that the compound turns into ether gas when left alone at room temperature. The word 'mole' has various meanings in colloquial English, ranging from a type of animal which lives in the ground, to a person who acts as an 'inactive' spy for the time being. The chemist, on the other hand, attaches a very different meaning to the term – which is actually derived from the word 'molecule'. We'll discuss the meaning of this word further below.

Superimposed upon this subtly changed 'normal' English there are words, phrases and concepts, and even ways of expression which were especially invented by the chemist for descriptive purposes. The ideal of the chemist is to find a way to give an unique name to each one of the millions of compounds, using a logical and general system of nomenclature which everyone understands and uses. Even a cursory examination of the 'old' names which have come down to us from the 'pre-scientific' periods of chemistry, shows that they are arbitrary and do not conform to any system of nomenclature based upon logic. Who, for instance, remembers the Mr Turnbull who proudly gave his name to the blue compound which he had discovered?

Chemistry has also become an *international* science. Scientists from all over the world who speak all kinds of different languages, want to communicate *unambiguously* with one anther. Chemical names, etc. should, therefore, be as similar as possible in the various main languages of the world. English has become the international language of chemistry and most of the nomenclature rules are based on the rules of grammar and spelling of the English language. It is, naturally, almost impossible to invent a set of rules for chemical nomenclature in English which will survive unchanged in any other language since the rules of spelling as well as grammar are generally different. However, it remains the ideal of the chemist to develop such a language of nomenclature.

The International Union of Pure and Applied Chemistry (known by the acronym IUPAC), is the international body responsible for nomenclature, notation, definitions, symbols and constants for the science of chemistry. IUPAC set itself the goal of designing a self-consistent language of chemistry based on a set of definitions of terms and concepts, as well as nomenclature. IUPAC has, therefore, published a series of nomenclature books, as well as books and recommendations dealing with definitions, symbols and conventions. These books form the definite rules which ought to be followed by every chemist. These rules are now followed by the majority of textbook writers. These

IUPAC books of nomenclature, etc. are colloquially referred to by the colours of their bindings, namely:

☐ the *Blue Book* (Organic Chemistry)[1]
☐ the *Green Book* (Physical Chemistry)[2]
☐ the *Red Book* (Inorganic Chemistry)[3]
☐ the *Orange Book* (Analytical Chemistry)[4]
☐ the *Purple Book* (Macromolecular Chemistry)[5]

These books are very complex and comprehensive and are, therefore, not easy to understand. Students of chemistry often experience difficulties comprehending the rules – even when they appear in a simplified form in an introductory textbook of chemistry. In this chapter we explain how English compound nouns, adjectives and even adverbs are built-up from simpler words. This serves as an introduction to the way in which the chemist builds up artificial names by the same process of word building, using a specially designed set of names (See Chapter 1: Words).

It is emphasised that we are not endeavouring to teach chemistry and chemical nomenclature in this chapter, but that we are presenting some aspects of the *language* of chemistry which forms a background to the study of chemistry. *This chapter, therefore, does not proceed along well-known chemical lines in which all chemical concepts are defined consecutively as though we were in a chemistry course, but we proceed in a rather 'strange' fashion, with emphasis on language and not on chemical concepts.* We draw the attention of the reader to aspects of the language of chemistry which will, hopefully, contribute to a better understanding of chemistry itself. The reader should not try to understand the limited amount of illustratory chemistry given in the text, but should concentrate upon the flow of the argument and upon the language itself. These examples only serve as illustrations for the use of language.

---

1 J Rigaudy and SP Klesney (Editors) *Nomenclature of organic chemistry – Sections A, B, C, D, E, F and H*. Published for IUPAC by Pergamon Press, Oxford, 1979.
2 I Mills, T Cvitas, K Homann, N Kallay and K Kuchitsu, (Editors), *Quantities units and symbols in physical chemistry*. Published for IUPAC by Blackwell Scientific Publications, Oxford, 1988.
3 GJ Leigh (Editor), *Nomenclature of inorganic chemistry*. Recommendations 1990. Published by Blackwell Scientific Publications for IUPAC, Oxford 1990.
4 H Freiser and G Nancollas (Editors), *Compendium of analytical nomenclature. Definitive rules*. Second Edition. Published by Blackwell Scientific Publications for IUPAC, Oxford, 1987.
5 W Metanomski (Editor), *Compendium of macromolecular nomenclature*, Published for IUPAC by Blackwell Scientific Publications, Oxford, 1991.

☐ Some Aspects of Lists and Tables of Classification

One of the most important things which must be remembered about chemists is their habits of *classification* and *systematisation*. They classify everything – elements, compounds, reactions, properties of matter, chemical structures, etc. – in their endeavours to understand matter in all its different forms, their properties and their reactions. They use many different ways of classifying and systematising, and the reader must be on the look-out for them. We'll give attention in this section to the search for common physical and/or chemical properties of elements or compounds, respectively. Such common properties and their variation within such a group or set of compounds, for instance, have special significance for the chemist and may help to predict the properties of new substances before they are even made.

Let us now make a few observations which may seem trivial, but which form the basis of all classifications in chemistry. Any set of objects (for example, the chemical elements) which has only *one property* each which differs from object to object can very easily be ordered according to the *numerical value of the property*. We usually use either a *vertical list* or a *horizontal list* to display the *names* of the set of such compounds, for instance, which are classified according to the physical property. For instance, we can place the name of the compound with the largest numerical value of the property at the top (or to the left) and that with the lowest numerical value at the bottom (or to the right), or *vice versa*. This type of ordered list is classified with respect to 'size'.

If the set of objects has *two properties or variables* each which can vary independently from object to object, then it becomes a bit more difficult, and a *table* is needed for a proper classification in which the numerical value of the one variable is ordered vertically and the other horizontally. Such tables can contain either a list of names or a list of names together with some numerical information. As an example, we give a brief discussion on the *periodic table of the elements* below (we'll also discuss other aspects of it later in the chapter).

About one hundred years ago chemists would have told us that this table was obtained by putting the names of all elements with similar chemical properties below one another, roughly following the order of the atomic 'weights' (which we don't define here). Today, we know that a list is obtained when we arrange the names of the elements in a row or column according to some property of the atoms of the elements which is called *atomic number* (see below). When we use a secondary classification parameter, namely the electronic structure of the atoms of the elements, then we rearrange the list of names of the elements into a table. Details of how the electronic structures of the atoms can be used to obtain this table are provided in chemistry

Table 11.1
The IUPAC thirty-two-column form of the Periodic Table

IUPAC 1988

Column	1	2														3	4	5	6	7	8	9	10	11	12	13	14	15	16	17	18
Row 1	1 H																														2 He
Row 2	3 Li	4 Be																									5 B	6 C	7 N	8 O	9 F	10 Ne
Row 3	11 Na	12 Mg																									13 Al	14 Si	15 P	16 S	17 Cl	18 Ar
Row 4	19 K	20 Ca														21 Sc	22 Ti	23 V	24 Cr	25 Mn	26 Fe	27 Co	28 Ni	29 Cu	30 Zn	31 Ga	32 Ge	33 As	34 Se	35 Br	36 Kr
Row 5	37 Rb	38 Sr														39 Y	40 Zr	41 Nb	42 Mo	43 Tc	44 Ru	45 Rh	46 Pd	47 Ag	48 Cd	49 In	50 Sn	51 Sb	52 Te	53 I	54 Xe
Row 6	55 Cs	56 Ba	57 La	58 Ce	59 Pr	60 Nd	61 Pm	62 Sm	63 Eu	64 Gd	65 Tb	66 Dy	67 Ho	68 Er	69 Tm	70 Yb	71 Lu	72 Hf	73 Ta	74 W	75 Re	76 Os	77 Ir	78 Pt	79 Au	80 Hg	81 Tl	82 Pb	83 Bi	84 Po	85 At	86 Rn
Row 7	87 Fr	88 Ra	89 Ac	90 Th	91 Pa	92 U	93 Np	94 Pu	95 Am	96 Cm	97 Bk	98 Cf	99 Es	100 Fm	101 Md	102 No	103 Lr														

The elements 57 to 70 are the rare earth elements or lanthanides (whose columns are not numbered) while elements 89 to 103 are called the actinides (whose columns are also not numbered). They are the so-called f-elements. This periodic table is the one adopted by IUPAC in 1988; some older books use periodic tables with different numbering schemes. s-electrons occur in the elements of columns 1 and 2, d-electrons in those of columns 3 to 12, and p-electrons in those of columns 13 to 18, while the elements in the columns without numbers have f-electrons.

courses. We want to emphasise here that we have *rows* in this table (they are sometimes called *periods*); the chemist looks at the variation of physical and chemical properties across such rows of the table. We also have *columns* in the table (they are also called *groups*) and they are numbered from 1 to 18, with a large gap in between (belonging to the so-called *transition elements*). We find, indeed, chemical similar elements in these groups, and chemists always search for the systematic change in chemical and physical properties down such a group. The periodic table is thus a very useful device for memorising such properties and how they change down a group or along a period of the table. We have always found it a great help to know the periodic table by heart, or, at least, large parts of it. See Table 11.1.

We now set some problems involving the use of such tables. The reader should ignore the fact that the questions sound 'chemical' and should concentrate only upon the language of the questions. The same kind of questions can be asked about any such table, and the key to answering them lies in finding the relevant position in the table which the question specifies. This position is then related to the information given by the statement in the question. The conclusion follows from understanding words such as 'increase', 'decrease', 'to the left', 'to the right', 'neighbour', 'diagonal', 'down', 'lower', 'higher', etc. – and involves no science at all!

Problem 1

If we say that some physical property (say, the electronegativities of the respective elements) decreases down a periodic group, would you expect the electronegativity of the element Fr (francium) to be lower or higher than that of the element just above it, namely, Cs (caesium)?

Problem 2

If we say that the numerical value of a certain physical property, (for example, the property called ionisation potential), increases going from left to right along a period of the table, would you expect that of Be (beryllium) to be higher or lower than that of B (boron)? If we then say that the ionisation potential also decreases down a periodic group, which of F (fluorine) and Cl (chlorine) will have the highest ionisation potential? (Hint: The gap in the table between Be and B in the second period does not really influence the validity of the statement.)

Problem 3

If we say that any element has some chemical properties in common with its neighbour which is one place to the right below it, which elements would be similar in this way to the elements K (potassium), Al (aluminium) and Be (beryllium) respectively?

Problem 4

If we say that those elements found to the left of the diagonal going through the elements Be, Al, Ge (germanium) Sb (antimony) and Po (polonium) are metals, would you say that Hg (mercury) is a metal or not? And At (astatine)? If we then say that the elements on the diagonal and their direct neighbours exhibit some properties which make them in-between being metals or non-metals, would you say that the elements As (arsenic), Sn (tin) and Tl (thallium) belong to this type?

Problem 5

If we say that all elements having d-electrons are called transitional elements, are Au (gold), Ti (titanium) and Mg (magnesium) transitional elements?

The smooth variation of one or more properties of a single substance or system with another physical property, is briefly dealt with in the chapter on the language of mathematics (Chapter 10), as well as in the chapter on the language of physics and physical chemistry (Chapter 12) under the heading *function*. Such a variation can be illustrated by another type of table which is discussed there. Some tables can also be nothing more than a collection of the same physical data for different elements or compounds, such as solubilities in water, melting points; such tables are called *tables of data*, and when they occur in the memory of a computer, they are called *data bases*.

The student should develop the ability to read and understand such lists and tables of numerical and other properties, since they form a part of the language and symbolism of chemistry.

☐ Chemical Elements

Let us look at an aspect of the classification system which underlies the entire subject of chemistry, namely, the way in which the chemical elements are named and classified. We'll explore the basis of the classification system while we are defining terms at the same time. To start off, we state the following:

> It has been experimentally established that there are just 103 different *chemical elements* which form the fundamental *chemical* building blocks of all matter in the universe (of which 90 occur in nature and 13 were made by man in nuclear experiments).

Although we are at this stage quite certain that we cannot find any more chemical elements in nature by means of our chemical analyses, there is a real possibility that more such synthetic elements may be discovered in giant 'atom smasher machines' (such as those at Geneva). This, however, does not detract from the validity of our definition of the concept *chemical element*, since the *full set* of elements represents the fundamental *chemical building blocks* of nature. It does not matter that one or more further man-made building blocks may be added at a later stage, since it does not change the principle at all. It is important to note that we have given a rather crude definition of the concept chemical element at this stage; this definition will be made more precise below.

It must be remembered that any *specific chemical element* is *not* a name, but a *piece of matter which is uncontaminated by anything else* – not even by another chemical element. This definition of the concept *chemical element* specifies the concept somewhat better than the one given above.

Each one of these chemical elements is given a *unique name* and associated with this name is a *unique abbreviation* which is called its *chemical symbol*. These names and their associated chemical symbols, which are recognized by IUPAC, must be learnt by heart; the more quickly this is done, the less traumatic is the study of chemistry.

Usually, a student is presented with a table (called the *periodic table of the elements* which we have already reproduced above) which displays all the chemical elements and their symbols in an *ordered and chemically sensible* way. It is, however, easier to learn these names and their corresponding chemical symbols from an alphabetical list and it is usually a good idea to make such an alphabetical list for oneself:

The unique *chemical symbol* which represents an element is either one capital letter (for instance, S, for sulphur), or a capital letter followed by a lower case letter (like As, for arsenic); no three-letter chemical symbols are used. Such a pair of letters in a chemical symbol for an element, such as Ti, Sc, Zn and Te always occur together as *one symbol* and can never be separated. The first letter in such a chemical symbol is always written with a capital letter and the second as a lowercase letter. The *full names* of the elements are never written with an initial capital letter, except when occurring at the beginning of a sentence.

There is no sign of any systematic nomenclature in this list of names of the elements and corresponding chemical symbols of the chemical elements. Some of the names come from antiquity (such as sulphur, aurum (gold), argentum (silver), and plumbum (lead)), whilst others are quite modern. Many modern names end with the suffix *-ium*. Some of the names of elements are derived from geographical names (for example, *rhenium* from the river Rhine), others from the names of great scientists (for example, *einsteinium* from Albert Einstein, *lawrencium* from Lawrence, the builder of the first cyclotron where some man-made elements were created), others from the names of well-known people (like *nobelium* from Nobel), others from the names of countries (like *polonium* from Poland, *germanium* from Germany, *francium* from France), yet others from the nickname of the devil, 'Nick' (*nickel*); from fairies or 'Kobolds' (*cobalt*); from the sea-God, Neptune (*neptunium*); or the God of the Underworld, Pluto (*plutonium*).

In ten cases the name of the chemical element from which its chemical symbol was derived differs from the modern English equivalent (for instance, W, *wolfram* for *tungsten*). These archaic names and symbols must simply be learnt by heart. The English equivalents for the names are used in the text of books, but the IUPAC symbols are used in chemical equations.

We have said above that the Universe is built up from combinations of *103 chemical elements*. We now list some things which we know about elements:

☐ From the definition of a chemical element, we know that any piece of matter can be separated in such a way that a sample of each constituent chemical element can be obtained *pure*.

☐ Every chemical element can be *distinguished* by chemical and physical tests from any other element. This is done in the branch of chemistry called *Analytical Chemistry*, and we say that we are doing *chemical analysis* (plural: analyses). There are two types, namely, *quantitative analysis* where we determine *how much* of a given element is present, and *qualitative analysis* where we determine which elements are present in a given sample of matter.

☐ Every chemical element is unique, but we find that some elements behave in a *similar fashion* under the same conditions of reaction. This gives us the opportunity to arrange the elements into some kind of scheme or table where we use the convention of placing similar elements in columns below one another. This concept of *similarity* is very important in chemistry and we find that chemists are always on the lookout for similar properties.

☐ Chemical elements can be mixed with one another, forming a *mixture*, in which the components do not react with one another.

☐ Chemical elements can react with one another, forming *chemical compounds* of which the properties are different from those of the constituent elements. These compounds occur everywhere in nature, and man has learnt to manipulate the reactions of elements, as well as the reactions of compounds with other compounds in such a fashion that we can *synthesise* new compounds. We also speak of the *synthesis* of compounds when we mean that we make them in the laboratory.

☐ Definitions in Chemistry

Definitions are very important in chemistry and every definition should not just be learnt by heart so that it can be reproduced in parrot fashion, but it should be *understood*. If a definition is understood, then it can be used with confidence when required. It is important to realise that *all definitions in chemistry are related to one another*. Any definition rests, as it were, on previous definitions. It is important to try to see the relationships between various definitions given in a chemistry course, to see how the one is built-up on the other, to see how alternative definition pathways are set up, to see the chain of definitions leading back to the fundamental laws of nature governing chemistry.

It must also be realised that all *definitions of chemistry, taken together, form a dictionary* of chemistry. In such a chemical dictionary we would find explanations (definitions) of all *chemical terms* (such as atom, element, particle), all *chemical verbs* (such as: mix together, precipitate, crystallise (out), dissolve, titrate), all *chemical adjectives* (such as: hydrochloric, benzenoid) and all *chemical adverbs* (such as: well (in 'a well-behaved reaction')). It is, therefore, a good idea to make one's own dictionary by writing down all definitions in sequence of occurrence (this list can be converted to a private 'dictionary' later by ordering the defined terms alphabetically, but it is not really neccessary, since the chapters of a textbook are often almost self-contained units; whenever cross-references to other definitions occur in later chapters, it is usually clear to which previous definitions they refer).

Not a single unneccessary word is used in definitions: all words are important and their precise chemical meanings must be taken into consideration when a definition is analysed. Definitions are very compact and precise; they serve to define any term or name or concept exactly so that there can be no possibility of any misunderstanding.

Attention is drawn to the fact that not all definitions in chemical textbooks are given in special sentences which are displayed in a different printing type and surrounded by a white border. Some 'less-important' definitions and explanations of terms and symbols used, are often found tucked away in paragraphs of text. They are also important and should be noticed, understood and remembered.

When a new definition is encountered, the meaning of every ordinary word and all chemical terms occurring in the 'explanation part' should be found. A dictionary should be used to obtain the meaning of all unknown 'non-scientific' terms. Any own 'chemical' dictionary of previously encountered definitions can also be used to obtain the meaning of some of the scientific terms used in a definition. If the meaning of a concept or word is not found, then another textbook may be helpful (library!), or the lecturer may be consulted. The meaning of all words occurring in a definition must be understood before the definition as a whole can be understood.

Once a definition is understood, it is a good idea to see just how it is related to previous definitions and how it is used in the text which follows: no definition is unnecessary and it will be used in the textbook. It might even be used in other chapters of the textbook!

Let us consider some examples of interrelated definitions. Since these concepts form the basis of much of the fundamental theory of chemistry they should be well understood. We concentrate upon their language, how they are formulated and upon the underlying philosophy – not upon their chemical content. All definitions should be analysed in the way shown below so that their relationships may be determined.

Definitions

The IUPAC definitions which we want to consider are those defining the fundamental concepts *atom, atomic number of an atom* and *chemical element*:

☐ An *atom* is the smallest unit quantity of an element that is capable of existence whether alone or in combination with other atoms of the same element or of other elements.

☐ The *atomic number* of an atom is the number of electronic units of positive charge carried by the nucleus of that atom.

☐ An *element* is matter, all of whose atoms are alike in having the same positive charge on the nucleus.

The first thing to notice about these definitions is that they all have the same form, namely, they are all *statements* of the type:

(Term or concept) is (description). (See Chapter 6).

The (*term or concept*) is the thing which is defined (atom, atomic number, element). The part of the definitions which we have indicated by the word (*description*) contains the actual definition or explanation of the concept. This part can use only *concepts which were defined previously* to explain the new concept. This part links the new definition to previous definitions, including those found in dictionaries of the English language, as well as the meanings of those 'ordinary' words to which the chemist attaches specialised meanings.

Let us now look at the first definition above. It gives an exact explanation of the term *atom*, a term which was used for the first time by some Greek philosophers about 2400 years ago. Its original meaning refers to the fact that these philosophers asked themselves the questions: 'What happens when we take a knife and cut a piece of matter in half, then one of the halves in half, and so on? Can we continue cutting a half in half forever, or is there a point where we'll be obliged to stop?' Some of them felt that they would eventually reach a point where the piece of matter cannot be further subdivided. The Greek word used for this process of cutting a piece of matter is *atomos*, which is the verbal adjective (participle) of the Greek verb *to cut* which means something like *cut* or *cutting*. We use the truncated (or shortened) form today as the word *atom*. The definition found in dictionaries for the word *atom* is usually something like: *a particle of matter so small that it cannot be further subdivided*. It is clear that IUPAC has given a much more precise definition. The words *the smallest unit quantity of an element capable of existence* form the part which we called the 'description' of the definition. This is the central part of the definition. The parts *alone* and *in combination with other atoms of the same kind or of other elements* merely qualify the word 'existence'.

When we read the definition, it follows that we first have to know what an *element* is in order to understand it. Some definition of the term element must be found before the meaning of the definition can be understood; tracing the path back to the empirical definition of the concept element would take us too far into the subject of chemistry itself and we assume that the reader will find it. The definition tells us that there are things in nature called *elements* which consist of *atoms*. (Which part of the statement tells us that there are more than two elements?) It also tells us that an atom of an element can exist ('occur') in three different ways in nature:

☐ It can exist on its own.
☐ It can combine with other atoms of the same element.
☐ It can combine with atoms of one or more different elements.

It is left for the reader to determine which words in the definition tell us this!

We point out that the little word *smallest* (the superlative of the adjective *small*) plays an extremely important role in the defintion. If a particular object from a collection or set of objects is the *smallest*, then no other object in the collection may be smaller than it. This adjective of comparison, *smallest*, qualifies the term *unit quantity*, telling us that there are other unit quantities of matter which are used in chemistry (there are, for instance, the gram, the kilogram or the mole). However, there is only one which is the smallest, and that particular unit quantity is called an *atom*!

Problem 6

There are cases in nature where it is found that one atom of an element can combine with one atom of the same element or with one atom of another element. Is the definition of IUPAC for an atom correct for these cases? If it is found that it is indeed not precise enough to include this case, then reformulate the definition to include it.

The next definition, that of the term *atomic number*, rests upon the previous definition of the term *atom*.

Problem 7

Which other terms or concepts need be defined before we can attempt to understand the definition *of atomic number?*

The definition of the third term, *element*, rests upon the definition of the term *atomic number*, and thus eventually upon the definition of the term *atom*. It is thus clear that we have a chain of definitions here which depend upon previous definitions.

Problem 8

One of the prime requirements of a logically constructed system of definitions needed for a science is that its set of fundamental definitions must not contain cyclic definitions of the type *definition A implies knowledge of definition B*, as well as *definition of B requires knowledge of definition A*. Is there anything in the three consecutive IUPAC definitions given above which seems to be of cyclic nature?

Mole

We mentioned the word *mole* in the introduction to this chapter when we pointed out that some ordinary words have different meanings in chemistry. Let us give a definition of this term, which is one of the central terms in chemistry.[6]

We often ask questions like: 'How much matter is there in this piece of matter?' To answer it, we have to know how to measure the *amount of matter* in any particular sample of matter. Once we can answer this question, then we can, for instance, compare two pieces of matter.

---

6 In the sections below we give some definitions in the text without drawing special attention to them. The reader is urged to find them and to write them down on a piece of paper in order to keep track of the flow of the argument – which is only based upon common logic and arithmetic which is easy to follow.

On a *macroscopic scale* or real-life scale, this is not so difficult, since we make use of the *mass unit* which we have arbitrarily designed, namely, a block of pure platinum which is kept in a vault in Paris. The *Bureau International des Poids et Measures* (The International Bureau of Weights and Measures) has decreed that the amount of matter in this block of platinum is exactly *one kilogram,* and we say that it represents a *mass of exactly one kilogram.* The rest of the matter in the universe is measured with respect to this standard mass. We say, for instance, that an object has a mass of 100 kilogram meaning that it consists of an amount of matter which is exactly one hundred times more than that of our mass unit, namely the kilogram kept in the vault in Paris. The masses which we use are thus masses which are *relative to the standard mass,* the kilogram.

It is not as easy to find a way of expressing the mass of an object on atomic scale. We need a definition of a standard mass on atomic scale, so that we can express the masses of the atoms of all the elements relative to that chosen mass as unit.

IUPAC decided to use a certain kind of *carbon atom* (we are thus again back at our fundamental definition at the beginning!) to obtain such a unit and said that its mass is 12 mass units *exactly.* Once we accept this as the standard mass for our atomic masses, then we can measure the masses of all other atoms with respect to it. We call such masses *relative atomic masses,* where the adjective *relative* immediately tells us that it is relative to the mass of some *chosen* atom. The meaning and use of the adjective *relative* must be well understood, since it occurs in many technical chemical terms: whenever it occurs, we know that it must refer to some arbitrary chosen standard.

Before we leave the question of mass standards, we want to say that we need another definition to bridge the gap between the macroscopic mass unit, the kilogram, and the atomic unit of relative mass, namely the mass of carbon-12 (as it is called). We need to know how many atoms of carbon there are in a certain macroscopic mass of pure carbon. A block of pure carbon-12 weighing 12 gram exactly was chosen to be the block of matter which to furnish the bridge to the atomic mass standard. In this block of carbon-12 the chemist established that there is the so-called *Avogadro number* of atoms of carbon-12, namely, the number $L = 6.022 \times 10^{23}$ atoms; it should be noted that we say 'Avogadro number', but that IUPAC allows us to use either the symbol N_A or the symbol L. It now follows that we can calculate the real masses of the atoms themselves by dividing the mass 12 gram of carbon-12 by L; this gives a very small number which is in the order of 10^{-27} kilogram.

IUPAC says that if we have 6.022×10^{23} atoms (or molecules) of any substance, then we have *one mole* (plural: moles) of atoms. Twelve gram exactly of carbon-12 are thus one mole of carbon-12. The concept *mole* plays a very important role in chemistry. It is important to realise that 1 mole of anything discrete contains 6.022×10^{23} of such discrete particles (whether we refer to atoms, to molecules, or even to people[7]). The unit itself, which is abbreviated 'mol', is never written in italics, but always in upright type, and the plural of the unit is never indicated, that is, we write '2 mol' for two moles of a substance.

Augmented symbol

A symbol of an element can be augmented by four numbers which can be attached to it; these numbers indicate either specific physical properties of the atom of the element in question (of which exact definitions must exist), or how many of the atom type are present in a compound.

There are two *superscripts* or *upper indices*; one is written in smaller type just to the left of the top of the symbol (it is called a *left superscript* or a *left upper index*) and the other just to the right of the top of the chemical symbol (it is called a *right superscript* or *right upper index*), respectively.

There are two *subscripts* or *lower indices*; the first is written in smaller type just to the left of the 'foot' of the chemical symbol (it is called a *left subscript* or *left lower index*), while the other is written just to the right of the 'foot' of the symbol (it is called a *right subscript* or *right lower index*).

In elementary chemistry we normally use all four possibilities. These four positions are marked as follows, for instance, for the case of the sodium atom, Na:

left upper index   right upper index

Na

left lower index   right lower index

IUPAC has established the *convention* that we write the value of the atomic number, Z, of an element (if we need to indicate it) as a *left lower index before the chemical symbol* of an element X:

$$_Z X, \text{ for example, } _6 C$$

The sum of Z, the atomic number, and N (the number of particles called, neutrons, in the atom of a particular atom) is called the *mass number*, which is according to the conventions of IUPAC, written as a *left upper index* before the chemical symbol of an element X:

$$_Z^{Z+N} X \text{ for example: } _6^{12}C \quad _6^{13}C \quad _6^{14}C$$

---

7 This number is really very big – the population of China is considered to be a very large number, yet it is only about one thousand million or 1×10^9; we'll need the population of one hundred million million Chinas to give us one mole of people!

The reader is advised to turn back to the definition of an element given above and read it again, thinking of the three *isotopes* of carbon, all of which belong to the element carbon, and all of which are listed on the same block in the periodic table of the elements. The name isotope comes from the Greek word meaning 'in the same place' – they occupy the same place in the periodic table, even though the atoms are slightly different from one another.

The *electrical charge* on an atom (which is caused by an electrical imbalance between the number of elementary positive particles, called protons, and the number of elementary negative particles, called electrons, in an atom) is written as a *right upper superscript* behind the symbol of the atom, for instance: Na^+ and Cl^-. We call such a charged entity an *ion* (pronounced something like 'i-yin'). The positive ions are the *cations* (pronounced something like: 'cat-i-yin.'), while the negative ones are the *anions* (pronounced 'an-i-yin').

The *number of atoms of a specific element in a molecule* is indicated by putting the particular number as a *right lower subscript* behind the chemical symbol of the atom, for instance, Na_2SO_4. If need be, we can use any of the various combinations of indices to specify an atom more accurately, for instance

$$_6^{12}C^+.$$

☐ Chemical Formulae

We have said above that atoms of the elements can either react with one another or with atoms of other elements to form *compounds*. These reactions take place in a variety of ways and under many different reaction conditions. Over the centuries chemists have isolated a bewildering number of compounds. There are, for example, millions of different carbon compounds synthesised by the organic chemist. Each of these compounds was:

☐ *isolated*, which means 'obtained' by separating it from all the other products which may have formed during a reaction

☐ *purified*, which means that all contaminating substances were removed, for instance, by repeated dissolution and crystallisation operations

☐ *characterised*, which means that its *chemical formula* (a complex symbol summarising the number of atoms of each element which is present in the compound) was established and that some other *physical properties* (such as melting point, boiling point, etc.) were determined

☐ *named*, which means that an *IUPAC systematic name* was given to it according to the rules of 'grammar' for IUPAC nomenclature

☐ *chemically characterised*, which means that at least the most important of its *chemical properties* or *chemical reactions* it engages in were determined.

Many ways of writing down these chemical reactions and the products obtained, using a type of *symbolic language*, have developed over the years and have to be learnt; most of these writing conventions are also to be found in the IUPAC rules. The way of writing down formulae and reactions is just as much part of the language of chemistry as definitions and vocabulary, since a *reaction equation* (as it is called) is just a shorthand and symbolic way of writing down a whole sentence. One can ask the question whether this symbolic language is needed at all, whether the same information cannot just as clearly and comprehensibly be transmitted in an ordinary full sentence. The answer to this question is a definite 'no', since we are dealing with extremely complex 'objects' (such as chemical formulae) which are extremely difficult to describe briefly and comprehensively in full sentences.

We have already described the first step in this complex process of developing a symbolic language, namely the assignment of a unique name to every chemical element, as well as a unique symbol, usually derived from the name of the element. This chemical symbol of an element acts in two ways:

☐ it serves as an abbreviation in the text for the element itself (for instance: 'Take 0.12 g of Au and react it with *aqua regia* ...'), and

☐ it serves as a symbol which indicates that there is just one atom of the element gold present (for instance: 'One Au atom reacts with ...'.

The chemist has several ways of writing down the *chemical formula* of a compound:

☐ The *empirical formula* indicates only the *ratios of the elements present in the compound*, for instance, the compound benzene has the empirical formula CH, indicating that there is one carbon atom for every hydrogen atom present in the molecule.

☐ The *molecular formula* of the compound shows the *actual numbers of the atoms present in the compound*, for instance, benzene is written C_6H_6.

    – The value of the molecular formula is sometimes enhanced as it were, by breaking up the molecule into 'chemically meaningful parts' to tell the reader about certain important aspects which give an indication of the reactivity of the molecule, or to which type of compound it is classified, for instance: $CH_3 \cdot CH_2 \cdot COOH$, instead of $C_3H_6O_2$, to show that it has the acid group -COOH, or Pb(OH)Cl, instead of PbHClO, to show the presence of the hydroxide ion OH⁻ and the chloride ion Cl⁻. The information needed to write down the formula in this way can come from a knowledge of the chemical properties of the compound, from its physical properties, and from a knowledge of the next type of formula, the *structural formula*.

☐ The *structural or geometrical formula* shows the actual positions of the atoms of the compound in space. Instead of giving sets of atomic co-ordinates in three dimensions, the chemist traditionally identifies *chemical bonds* in the compound, that is, which atoms are linked to which atoms by means of electrical forces operating within the compound. The structure of the compound is then represented by a drawing of the compound in three dimensions, showing the *lengths of these bonds*, as well as the *interbond angles* and *interplane twist angles* (see below).

☐ The *stereoscopic formula*, where the structure of the compound is drawn to scale by a computer in such a fashion that it can be viewed in stereo, using a simple stereoscope. This type of structural entity can give a breathtaking view of the way a compound is constructed.

Compounds are isolated in the form of all four main phases of matter, namely, the *gaseous phase*, the *liquid phase*, the *solid phase* and the *amorphous phase*. The noun *phase* simply means a well-characterised way in which matter can exist. The adjective *amorphous* means that the compound seems solid, but that it does not have a discernible structure like a crystal and is actually without form or consistent internal structure.

Some compounds consist of one single, gigantic structure, for example, a crystal of carbon, called diamond (which consists of billions of carbon atoms linked together in an unchanging pattern) or a crystal of table salt, called sodium chloride (which consists of sodium cations and chlorine anions alternating in the three directions of space). In such cases, the empirical formula is also the formula of the compound itself.

Other compounds consist of isolated units, called *molecules*, which can exist on their own and which clearly show up in the structural study. In such cases, the formula of the compound, is then called the *molecular formula*. The *geometric* or *structural formula* shows the actual dimensions of the molecule.

Figure 11.1
A simple three-
dimensional
molecule

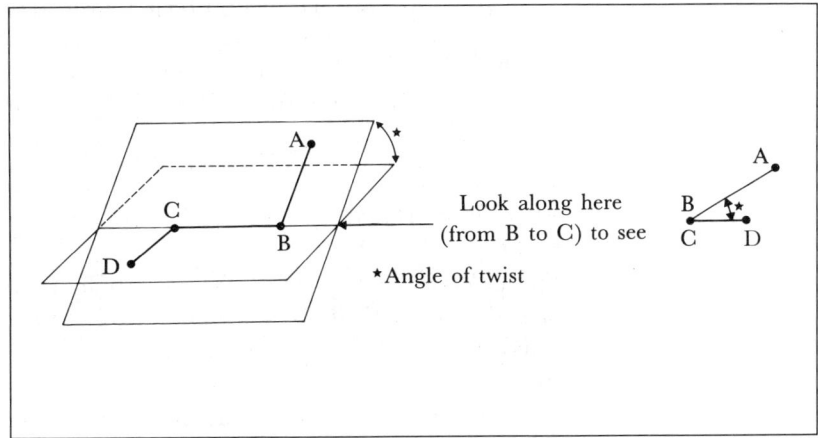

A simple three-dimensional molecule is depicted in **Figure 11.1.** There are three different *bond lengths*, namely r_{AB}, r_{BC} and r_{CD}. There are just two different *bond angles*, namely ABC and BCD, while the *angle of twist* between bonds AB and CD (in that order) is shown in perspective and in side-view. The chemist places great value upon the knowledge of these three *molecular parameters* for a molecule or ion which is under consideration, since they can tell him very much about the bonding situation in the molecule and about the arrangement of the electrons in a molecule, and, hence, about the reactions the molecule or ion can engage in. Tables of bond lengths and interbond angles are usually found in every elementary textbook of chemistry and are used for purposes of comparison or analysis of trends. The trends exhibited by the data in these tables are usually of the utmost importance.

☐ Reaction Equations

Instead of saying that lead nitrate reacts with sodium chloride in cold solution to give a precipitate of lead chloride and a solution of sodium nitrate, we use a symbolic short-hand, and write down the chemical formulae of the reactants, as well as the products. The way of writing is as follows:

$$Pb(NO_3)_2 \text{ (aq)} + NaCl \text{ (aq)} \dashrightarrow PbCl_2 \text{ (s)} + NaNO_3 \text{ (aq)}$$

It is immediately seen that this writing convention is a very powerful way to contract a long sentence into a short reaction equation. This means that equations like this must be very carefully read, since they convey a great deal of information.

Various abbreviations are used in the above equation, such as:

☐ The *abbreviation* '(aq)' immediately tells us that the reaction takes place in *aqueous* solution (water is *aqua* in Latin). Other *solvents* (other liquids which *dissolve* a compound; the dissolved compound in the solution is called the *solute*) are known, such as a number of organic solvents; some have separate abbreviations, but the presence of others is indicated by writing the name down over the reaction arrow.

☐ The *abbreviation* '(s)' tells us that $PbCl_2$ is a solid and that it 'precipitates out' of the solution. If a gas were present during a reaction it is indicated by the sign (g), and in the case of a separate liquid, it is indicated by (1).

☐ The *reaction arrow* tells us that the reaction takes place from left (where the *reactants* are) to right (where the *products* are).

  ☐ If the arrow *points to the left*, then the reactants are on the right and the products on the left and the reaction takes place 'backwards'.

  ☐ If there is a *double arrow* ($\rightleftharpoons$), it immediately tells us that the reaction can proceed both ways and that a *reaction equilibrium* will always be established, where we'll always obtain all four compounds (in this case) in constant mass ratios in the solution. The lengths of the double arrows sometimes differ ($\rightleftharpoons$ or $\rightleftharpoons$), showing that the equilibrium is biased towards the side of the reaction to which the longer of the two arrows is pointing.

☐ The *addition sign* ' + ' on the side of the reactants is read as *reacts with* (if there are more than one such ' + ' signs on the left, then the second and further signs are read as *and*). This sign occurring on the side of the products is simply read as *and*.

When the above equation is carefully studied, it is seen that the numbers of the atoms of the different elements do not agree on both sides of the arrow. The reaction arrow can be replaced by an equality sign ' = ' when the numbers of atoms agree on both sides. Such an equation is known as a *balanced equation*. It is easy to balance the above equation by inspection:

$$Pb(NO_3)_2 \text{ (aq)} + 2 \text{ NaCl (aq)} = PbCl_2 \text{ (s)} + 2 \text{ NaNO}_3 \text{ (aq)}$$

How this balancing is done, is taught in chemistry courses. The equal sign can only be used if all the number of atoms balance on both sides. Balancing a chemical equation can only be done when all the products have been identified. Balancing a chemical equation forms part of a very important chain of operations which makes chemistry a quantitative science. When a chemical equation is balanced, one can then write it in terms of the moles of reactants and products:

$$Pb(NO_3)_2 \text{ (aq)} + 2\ NaCl \text{ (aq)} = PbCl_2 \text{ (s)} + 2\ NaNO_3 \text{ (aq)}$$

1 mol   2 mol   1 mol   2 mol

In this case, the number of moles on both sides of the equal sign also balance, but it is not a necessary condition for the number of moles to balance; for instance, it does not happen in the case of the reaction

$$C \text{ (s)} + O_2 \text{ (g)} = CO_2 \text{ (g)}$$

1 mol   1 mol   1 mol

Once we have the number of moles for a balanced chemical equation, we can relate it to the *masses of reacting species and products* by means of the individual *molar* masses. For instance, we know that the molar masses of carbon is 12.01 g, that of molecular oxygen is 31.998 g, while that of carbon dioxide is 43.998 g. This means that carbon and oxygen will always react in these ratios of masses, giving the same mass ratio of carbon dioxide. Balancing an equation is thus a neccessary step in chemistry.

☐ General Principles of Systematic Nomenclature

It is important to realise that any systematic chemical name is not a simple noun with which we are familiar in ordinary English, such as *tree*, or *Jack*. A systematic name in chemistry is built up in a logical fashion and there are different ways of generating a name of a compound. We choose to describe the process of giving a name from the point of view of the formal grammar of the English language, but we do not neglect to indicate some of the ways which IUPAC uses.

We start our introduction by looking at the use of the *normal combination of adjective which modifies a noun* described in Chapter 1. We then describe the use of *nouns as adjectives to modify other nouns*; this type of name is used quite often in chemistry. This is followed by a description of the construction of *compound nouns and adjectives* and their rules of spelling. These concepts are then applied to analyse chemical names which were generated by the IUPAC nomenclature system.

The first problem to be solved in the naming of chemical compounds is to find the *main noun* of the name. Here the chemist makes use of the custom of classifying compounds into similar types to find the main nouns of the names, such as, for instance, the collective nouns *acid*, *hydroxide*, and even parts of names which function as names of compound types, for instance, the suffixes *-ane* (which is derived from the set of compounds collectively called the *alkanes* and which functions as a type of noun), and *-ol* (which comes from the set of compounds collectively called the *alcohols*).

The other main difference between the naming of chemical compounds and that which occurs in non-chemical English is found in the construction of *molecular naming fragments* which are actually nothing more than frequently-occurring parts of molecules. The names of these molecular naming fragments are then 'stitched' to the main nouns in a systematic fashion in several ways.

Some other differences between chemical nomenclature and ordinary word-building are as follows:

☐ The occurrence of various types of *brackets* like the bracket pairs ( ), [ ] and { } in names (and also in formulae); they are used to isolate certain *parts of words* which function as sub-naming units.

☐ The use of numbers and other symbols (such as Greek letters), which are called *locants* which indicate where the naming units are attached to the central backbone naming unit.

☐ The conventions for naming the positions of the atoms in the three-dimensional space of a molecule. Words like *cis* (from the Latin for 'on this side'), *trans* (from the Latin for across), and a system of assigning letters to name the corners, edges and sides of geometrical figures in three dimensions.

There are really two nomenclature problems which are inverses of one another. The first is the construction of the systematic name from the molecular or structural formula (as the case may be). The second is the reconstruction of the molecular or structural formula from the systematic name of the compound. Both make use of the same set of rules as developed by IUPAC. It is thus neccessary to understand these rules and their application very well so that both the direct, as well as the inverse pathway may be used with confidence.

Nouns used as adjectives

Our analysis starts off by noting that it is possible to use a noun as an adjective to describe another noun in English. Some examples of nouns used as adjectives plus their following nouns are given below, where the main noun is printed in bold:

☐ the glass **flask**

☐ the chemistry **laboratory**.

The nouns which are used as adjectives are *glass* and *chemistry*. It is also possible to have more than one noun used as adjectives, for instance, *New York State Police Force Pension* Fund.

It is to be noted that both parts of the expression, namely the noun(s) used as adjective(s) and the following noun, are equally emphasised. This makes this usage different from that of the following type, namely, the *compound word*, where the whole is more than just the sum of the parts.

Compound nouns and adjectives

The meaning of a *compound word* is different from the meaning of its parts, for instance, many letters are found in boxes, but a letter-box is put on street corners to post one's letters in; there may be many a mother who is a hood (or a crook), but where the suffix -*hood* is originally attached by a hyphen to the word mother to give *motherhood*, the meaning is very different; and there are many people who do not do well, but a *ne'er-do-well* has a very particular meaning. In English such a compound word is usually *hyphenated* when it is used for the first time; the hyphen sometimes disappears with time (for example, *goldfish, quicksand, flyover,* or even *motherhood*).

There are many ways in which these compound words are formed in the English language; the most important from the point of view of chemistry, are the following:

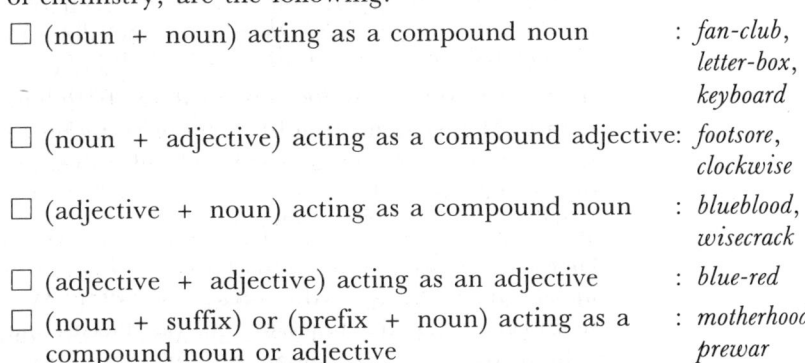

☐ (noun + noun) acting as a compound noun   : *fan-club, letter-box, keyboard*

☐ (noun + adjective) acting as a compound adjective: *footsore, clockwise*

☐ (adjective + noun) acting as a compound noun   : *blueblood, wisecrack*

☐ (adjective + adjective) acting as an adjective   : *blue-red*

☐ (noun + suffix) or (prefix + noun) acting as a compound noun or adjective   : *motherhood prewar*

General ways to generate chemical names

There are the following major ways in which systematic chemical names are constructed using the four general types of grammatical building blocks which we discussed above, namely:

(1) simple noun

(2) simple noun acting as an adjective, followed by a simple noun

(3) compound noun

(4) simple or compound noun acting as an adjective, followed by either a compound noun or a simple noun

(5) compound adjective followed by a simple or compound noun.

Since the chemist is not in the habit of thinking in terms of such grammatical concepts as adjectives and compound nouns, etc., the description of chemical nomenclature as given by IUPAC bears little resemblance to the grammatical classification given above, but the two are basically the same. We'll continue with the grammatical analysis since the patterns then become very clear.

The compound nouns and compound adjectives of chemistry may be very complex indeed and may eventually reach several hundreds of letters and signs just for one word (the record is over eight hundred letters!). It is clear that one has to understand the grammatical construction underlying any such a 'big' word! The bridge to the names which IUPAC gives to these *grammatical* parts of chemical names will be given later in this chapter.

Before we proceed any further, we want to emphasise that molecules are, as it were, 'cut up' into *nomenclature entities or fragments* for purposes of nomenclature, each one having its own unique 'name'. These nomenclature fragments bear no relationship with what actually happens during the course of a chemical reaction (although there may be some reaction entities floating around during the course of a reaction which may look identical to the nomenclature fragments). These nomenclature fragments must be clearly seen for what they are: they are there to give systematic names to compounds by allowing us mentally to cut up the molecular formula into parts which can then be identified with some of these nomenclature fragments and their simple names. These naming fragments are then stuck together in a systematic fashion to form the compound word which then functions as the name of the substance.

We want to point out that, from now on, we refer to complex formulae by first displaying a molecular or structural formula which is then identified by a Roman numeral such as (XXI). We refer to these numbers in the text instead of using the name of the compound; the reader is expected to search for the formula every time it is encountered. This method is used very often since it saves space. It is also convenient because a chemical name is often very long and not easy to comprehend at a glance (with practice, a chemical structural formula can be very quickly comprehended).

(1) Simple nouns

These nouns are mostly derived from archaic nomenclature and are incorporated into IUPAC systematic nomenclature. Examples are common, such as *benzene* (I), *oxide* (a name which refers to the oxygen compounds of elements), *ion* (a name which refers to a positively or negatively charged molecular species), *cation* (referring to a positive ion, such as Na^+), *anion* (referring to a negatively laden ion, such as Cl^-), *nitrate* (referring to the ion NO_3^-), etc.

IUPAC very cleverly made use of these archaic names in the development of systematic nomenclature. For instance, the series of organic compounds known by the 19th century names of *methane* (CH_4), *ethane* (C_2H_6), *propane* (C_3H_8), *butane* (C_4H_{10}), *pentane* (C_5H_{12}), ..., are called a *homologous series*, since the molecular formulae of the successive members differ by a constant grouping of atoms (in this case it is $-CH_2-$). IUPAC realised that the names of all these compounds end with the common letters -ane, to which are appended the respective identifying parts meth-, eth-, prop-, but-, pent-, etc. IUPAC thus decided to use these parts as building blocks to form other nouns, having the noun -ane as 'end noun' or suffix as described above; we'll return to this below.

(2) Simple noun acting as an adjective, followed by its simple noun

IUPAC again made use of the archaic simple names of the type discussed in (1) above and incorporated them into systematic nomenclature. It is important to note that *such names are never written with a hyphen between the two nouns*. This type of name is especially used for certain types of inorganic compounds, such as those referred to as being classified in the sets of compounds collectively known as ion, acid, chloride, bromide, hydroxide, nitrite, etc. Examples of such names are, for instance:

> *sodium chloride* ($NaCl$), *sodium bromide* ($NaBr$), *sodium hydroxide* ($NaOH$), *lead bromide hydroxide* [$PbBr(OH)$].

It is customary to write the metallic part first in such names (where applicable), but the rest is not subjected to any order (for instance, the name *lead hydroxide bromide* is also allowed). There are certain rules, however, but discussing these will take us too far into chemistry.

It is important to realise that these nouns and their chemical representations (formulae, which include the signs of any charged species) must simply be learnt by heart. There is no other way to learn them, and the quicker this is done, the easier chemistry becomes. It is equivalent to learning the words of a new language: one should be able to know them even in one's sleep!

(3) Compound noun

There are several ways in which these compound nouns are generated, depending upon the branch of chemistry. We can only touch upon the subject here, presenting some of the most important ones.

There are indicators of number, that is, how many of the particular atomic, molecular or ionic entity are present. They are, respectively, indicated by the numerical prefixes, such as

---

☐ *mono-* (used only when it is absolutely necessary to indicate the fact that something occurs singly, for instance, *monosodium glutamate*)

☐ *di-* (for two, for instance, *dinitrogen oxide*, N_2O; *nitrogen dioxide*, NO_2; *disulphur dichloride*, S_2Cl_2)

continued

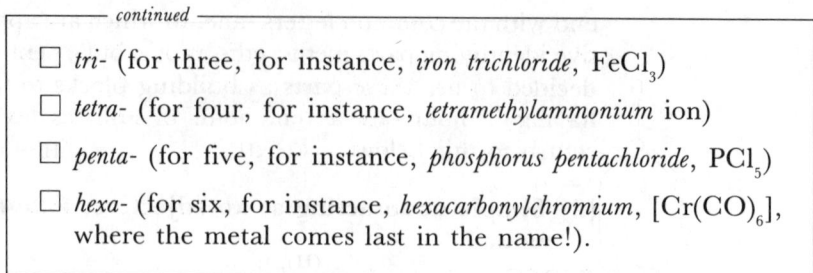

────── *continued* ──────

☐ *tri-* (for three, for instance, *iron trichloride*, $FeCl_3$)

☐ *tetra-* (for four, for instance, *tetramethylammonium* ion)

☐ *penta-* (for five, for instance, *phosphorus pentachloride*, PCl_5)

☐ *hexa-* (for six, for instance, *hexacarbonylchromium*, $[Cr(CO)_6]$, where the metal comes last in the name!).

There are further numerical prefixes for numbers greater than six, but we do not list them here. We emphasise that *the numerical prefixes are never written with a hyphen within a word*.

The chemist also prefers to use archaic names in cases such as Na_2SO_4, where we rather write *sodium sulphate* instead of the more systematic name *disodium sulphate* because every chemist knows that the sulphate group will attract two positive ions like the sodium ion; there is nothing, however, which prevents the use of the full systematic name.

Organic chemistry (the branch of chemistry which deals mainly with carbon compounds) was the first to develop systematic nomenclature. The method of nomenclature as developed by IUPAC allows the use of several nomenclature models. We do not discriminate between these models: we approach the IUPAC rules from the point of view of language. The word-types which we discuss here are thus applicable to all the possible models.

Let us look at the way in which one type of compound noun is constructed. We have already indicated the basis of the method when we discussed the concept of a homologous series. The organic chemist has learnt from experience that organic compounds can be classified in types, depending upon their chemical properties, which in turn depend upon several factors, such as the number of carbon atoms in the carbon skeleton (or carbon chain) of the molecule, the presence of certain groups (for instance, the groups $-OH$, $-CHO$, $=CO$, $-CH_2 = CH_2 -$, etc.), the spatial arrangement of the atoms (called the geometrical configuration), etc.

Let us consider the case of the homologous series of compounds which we discussed above. These compounds were called the *saturated paraffins* in the 19th century (saturated, because there are no double bonds in the molecule; paraffin, because they are the constituent molecules in the original household paraffin; they also occur in modern petrol mixtures). They are today known as the *saturated hydrocarbons*, since they contain only the elements hydrogen and carbon and do not have any double or triple bonds (they are, hence, *saturated*).

```
        H
        |
        C
      //  \
H — C      C — H
    |       |
H — C      C — H
      \\  //
        C
        |
        H
```

(I)

benzene C_6H_6

```
     H
     |
H — C — H
     |
     H
```

(II)

methane CH_4

```
    H   H
    |   |
H — C — C — H
    |   |
    H   H
```

(III)

ethane $CH_3 \cdot CH_3$

```
    H   H   H
    |   |   |
H — C — C — C — H
    |   |   |
    H   H   H
```

(IV)

propane $CH_3 \cdot CH_2 \cdot CH_3$

```
    H   H   H   H
    |   |   |   |
H — C — C — C — C — H
    |   |   |   |
    H   H   H   H
```

(V)

butane $CH_3 \cdot CH_2 \cdot CH_2 \cdot CH_3$

```
    H   H   H   H   H
    |   |   |   |   |
H — C — C — C — C — C — H
    |   |   |   |   |
    H   H   H   H   H
```

(VI)

pentane $CH_3 \cdot CH_2 \cdot CH_2 \cdot CH_2 \cdot CH_3$

```
    H   H   H   H   H   H
    |   |   |   |   |   |
H — C — C — C — C — C — C — H
    |   |   |   |   |   |
    H   H   H   H   H   H
```

(VII)

hexane $CH_3 \cdot CH_2 \cdot CH_2 \cdot CH_2 \cdot CH_2 \cdot CH_3$

```
     H
     |
H — C —
     |
     H
```

(VIII)

methyl- or $CH_3 -$

```
    H   H
    |   |
H — C — C —
    |   |
    H   H
```

(IX)

ethyl- or $CH_3 \cdot CH_2 -$

```
    H   H   H   H   H
    |   |   |   |   |
H — C — C — C — C — C — H
    |   |   |   |   |
    H  H—C—H H   H   H
        |
        H
```

(X)

2-methylpentane $CH_3 \cdot CH(CH_3) \cdot CH_2 \cdot CH_2 \cdot CH_3$

— continued —

---

continued ────────

$$CH_3 \qquad\qquad CH_3 \qquad\qquad\qquad CH_3 \qquad\qquad CH_2-CH_3$$

$$CH_3CH_2CH \;-\; CH_2 \;-\; CH \;-\; CH_2 \quad CH_2 \;-\; CH \;-\; CH_2 \;-\; CHCH_2CH_3$$

13 12 11   10   9   8 \ 7 / 6   5   4   3 2 1

C

/ \

$$CH_3CH_2CH \;-\; CH_2 \;-\; CH \;-\; CH_2 \quad CH_2 \;-\; CH \;-\; CH_2 \;-\; CHCH_2CH_2$$

$$CH_2 \qquad\qquad\qquad CH_3 \qquad\qquad\qquad CH_3 \qquad\qquad CH_3$$

(XI)

7,7-Bis(2,4-dimethylhexyl)-3-ethyl-5,9,11-trimethyltridecane

The molecular formulae of the first few members are given below: methane (II), ethane (III), propane (IV), butane (V), pentane (VI) and hexane (VII). When we look at them, we can see that they are all similar. IUPAC decided that the noun -ane is the 'family name' of all such similar compounds, as we said above (it is called a suffix by IUPAC, since it is added at the end of the name). We then have to distinguish between the individual members of the series by finding 'Christian names', as it were. These parts which function as 'Christian names' have already been indicated above, and the first four members are meth-, eth-, prop-, but-, while the higher members are named systematically, using the classical counting prefixes with the final 'a' deleted, namely, pent- and hex-, etc.

The names methane, ethane, propane, butane, pentane and hexane, respectively, refer to entities having 1, 2, 3, 4, 5, 6, etc. carbon atoms in a row together with their attached hydrogen atom atoms. This situation is shown in structures (II) to (VII). These are the names of the *parent hydrocarbons* which contain only hydrogen and carbon and are saturated; the names of all other non-cyclic saturated hydrocarbons are derived from the names of this homologous series.

This process of joining words to form new words is quite similar to that of non-chemical English as explained above. For instance, the word-building found in the hydrocarbons is similar to that found in the series of words: *widowhood, statehood, motherhood,* etc., where the suffix *-hood* plays the same role as the suffix *-ane*. Let us repeat the list of chemical names for the sake of clarity:

meth-ane   methane
eth-ane   ethane
but-ane   butane

prop-ane   propane
pentane   pentane
hex-ane   hexane.

Wherever the ending *-ane* is encountered, we know that the molecule belongs to the hydrocarbon type of molecule, no matter how complex it may look.

The IUPAC system now makes use of the principle of substitution, by mentally removing an end hydrogen atom from each of the molecules of the parent hydrocarbons to give a *molecular naming fragment* (which IUPAC calls *univalent radicals*). The fragment is indicated by deleting the -ane part of the name and adding the word -yl to the first part. In this way we have, for instance, methyl- (VIII), ethyl- (IX), etc. If we want to name the compound represented by the molecule (X), we say that it is a substituted pentane, since it contains five saturated carbon atoms chained together in an unbroken row. We say that the *substituent* methyl- (CH_3-) is found on the second carbon atom. The name is, hence, 2-methylpentane. The use of the hyphen should be noted. The numerical adjective 2 which qualifies the word methyl is called a locant by IUPAC, since it tells us where the methyl group or methyl radical (as it is called) is localised on the carbon skeleton. Let us repeat part of the list above to give a better impression of the method:

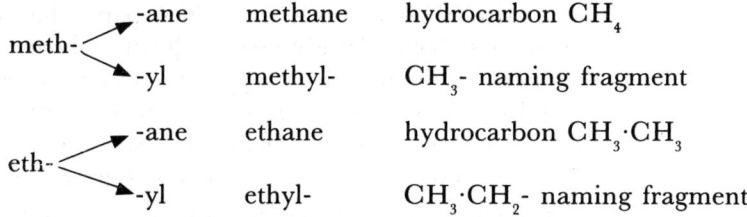

meth-⟨ -ane   methane   hydrocarbon CH_4
        -yl   methyl- CH_3- naming fragment

eth-⟨ -ane   ethane   hydrocarbon $CH_3 \cdot CH_3$
       -yl   ethyl- $CH_3 \cdot CH_2$- naming fragment

There are complex rules to determine which carbon skeleton (or carbon chain) should be used (there may be several possibilities) and from which side its carbon atoms should be numbered; these rules are found in chemistry textbooks. Many kinds of *chain substituents* are found in organic chemistry, such as bromo- (Br-), chloro- (Cl-), hydroxyl- (-OH), etc. It is now easy to imagine the formulae for compounds having names such as 2-hydroxypentane, 2-bromopentane, and 2,2-dichloropentane. The name 2,3-dichloropentane will not stretch the imagination too far.

The method is taken further by using other end names (they are called functional groups by IUPAC), such as -ol (for those molecules we identify as containing an end hydoxyl (-OH) group. This then gives, for instance, the name *pentanol*, which is obtained by leaving out the last -e of the name pentane to give pentan-, to which is added the name -ol. Full discussions of these rules are to be found in chemistry textbooks. Let us repeat part of the list above for the sake of clarity, adding the new step:

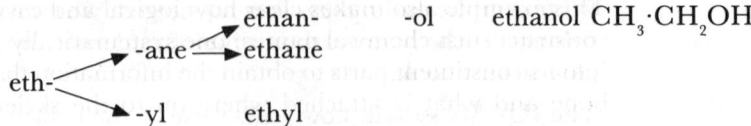

ethan-   -ol   ethanol $CH_3 \cdot CH_2OH$
-ane   ethane
eth-
-yl   ethyl

(4) A simple or compound noun acting adjectivally, followed by a simple or compound noun

The way in which these names are built up is now familiar, and we have names like *tetramethylammonium ion*, which is derived from the *ammonium ion*, NH_4^+, by methyl substitution of the hydrogens to give $[(CH_3)_4N]^+$. This name illustates the case of the compound noun used as adjective, followed by a simple noun.

The case of a simple noun used as adjective, followed by a compound noun is, for instance, represented by *potassium tetracyanonickelate*, $K_4[Ni(CN)_4]$.

(5) Compound adjective followed by a simple or compound noun

It is obvious that the case of the compound word used as an adjective, followed by a compound noun can also occur, for instance, *tetramethylammonium tetracyanonickelate*; we leave it to the reader to find the formula. We point out that the name is actually constructed from two names, namely *tetramethylammonium cation* and *tetracynanonickelate anion*, where the first words of each function formally as adjectives. The compound name is actually a contracted name, because the words *cation* and *anion* have been dropped for the sake of clarity. It is thus clear that this name also belongs to the type we identified above, namely, a noun used as an adjective.

(6) The use of brackets in names

The use of brackets is prescribed very precisely in chemical nomenclature and they function to remove any doubt which the reader may have about the position and nature of certain substituents on the molecular backbone. Let us look at the name of the compound 7,7-Bis(2,4-dimethylhexyl)-3-ethyl-5,9,11-trimethyltridecane. The name tridecane immediately tells us that the carbon skeleton of the molecule has 13 atoms; that there are three methyl groups situated on carbons 5, 9, and 11; that there is one ethyl group situated on carbon number 3, that there are two (the word Bis- tells us this) groups called 2,4-dimethylhexyl- in the molecule and that they are both attached to carbon atom 7. There are two methyl groups situated on the 2nd and the 4th carbon atoms of the hexyl group. The molecule is shown in structure (XI). The brackets are thus used to tell the reader about the two dimethylhexyl groups in the molecule; if they were not there, one would experience difficulties in deciding just where the groups should go.

This example also makes clear how logical and easy it is to read and construct such chemical names: one systematically dissects the name into its constituent parts to obtain the information that forms the backbone and what is attached where on to the skeleton. The reverse process is also easy, since the process is reversed: one determines the backbone, gives the correct name to it, finds the substituents on the chain, numbers the chain and then constructs the name from the fragments and their locants, keeping the use of hyphens and brackets in mind. All this presupposes that the names, the names of the naming fragments and the rules for finding the backbones of molecules are known!

☐ Answers

Problem 1

The word *decrease* means 'becomes less or having a lower value', that is, the numerical value of the physical property (electronegativity, in this case) is lower for each successive element of the first periodic group Li, Na, K, Rb, Cs and Fr. This means that the melting point of Fr must be lower than that of Cs, since it occurs under it in the group.

Problem 2

If some property *increases* then it becomes bigger and bigger. If the ionisation potential increases from left to right along a period, then it means that the ionisation potential of Be must be lower than that of B. The ionisation potential of F would be higher than that of Cl, since it is above it in the group.

Problem 3

We must go one row down from each of the elements and then move one place to the right. This gives us the similar pairs K and Sr (strontium); Al and Ge (germanium); and Be and Al. It should be noticed that the gaps in the second and third rows do not really count, since the instruction is clear: one period down, one position to the right.

Problem 4

Hg occurs to the left of the diagonal, so according to the defintion it is a metal, while At occurs to the right of the diagonal and is thus a non-metal. Looking at the table, we see that Sn and As are direct neighbours of the diagonal elements Ge and Sn; they, therefore, belong to this class. The element Tl is not a direct neighbour of any one of the diagonal elements, and is, therefore not of this type.

Problem 5

The lower part of the table gives the electron 'names' for each of the periodic groups. From this it follows that Au and Ti belong to groups which have d-electrons; they are, therefore, called transitional elements. When we look at the group in which the element Mg occurs (it is in the second column), we see that the electrons are called 's' electrons; it is, therefore, not a transitional element, since its electrons are not called 'd'-electrons.

Problem 6

The definition is indeed too restrictive because of the plural *other atoms* in the part ... *or in combination with other atoms of the same element or of other elements*. This plural implies that one atom of the first element can only combine with two or more atoms of the same element, or with two or more atoms of another element. The correct statement is something like: ... *or in combination with one atom or more atoms of the same element or other elements*.

Problem 7

To understand this definition we need to know terms such as *electronic units of positive charge* and *nucleus of an atom*, that is, we must know the definitions relating to the sub-atomic structure of an atom. When such defintions are encountered in a chemistry course these other definitions should be obtained.

Problem 8

There is a rather important apparent 'difficulty' in this little set of three definitions, since the first one, namely, the one of the concept *atom*, requires knowledge of the concept *element*, which is only formally defined in the third definition! This seems to be a set of definitions which is of cyclic nature. The solution to the apparent problem lies in the meaning of the term *element* which occurs in the first definition. The term *element* occurring in the first definition rests very heavily upon the empirical Greek 'definition' which was given earlier, as well as upon our collective chemical 'experience' of hundreds of years: we can *chemically identify* only 103 different fundamental building-blocks of matter, called elements. The term *element* which is used in the first definition *is made more precise* in the third definition by means of the second definition which tells us what *an atomic number* is. Science often makes use of such progressive definitions in which the meaning of a term is refined step by step.

12

An Introduction to the Language of Physics and Physical Chemistry

☐ Introduction

We give an introduction to the language of physics and physical chemistry in this chapter and emphasise some elementary aspects. Physics and physical chemistry are closely-related and the elementary aspects of their language can be treated in the same chapter. In addition, they are both based upon mathematics. In fact, it is impossible to study physics and physical chemistry without a sound knowledge of mathematics. We are thus not going to repeat what we have already said about the elementary aspects of the language of mathematics which were treated in Chapter 10.

We draw the attention of the reader to the following usage of the word 'physics' in this book: When the word 'physics' occurs alone in a sentence, it functions as an abbreviation for the words 'physics and physical chemistry' since the two subjects and the aspects of their language in which we are interested are very similar.

In this chapter we use some examples drawn from physics and physical chemistry. However, seeing as we concentrate upon the language (such as definitions, dimensions, units and visual information) – it is not necessary to understand the physics of the illustratory examples. We expect the reader to concentrate upon the logic of the argument and the flow of language as exemplified in the examples. We, therefore, urge the reader not to be side-tracked by trying to learn something about the theory of physics itself in this chapter.

Laws

Physics rests upon concepts which are called *laws*. These laws are primary assumptions made by scientists and they cannot be derived from any other theory: they are presumed to be true (till proved otherwise). The description of all other natural phenomena is given in terms of theories which are derived from these laws. Most laws come from our 'everyday experience' or from experiments performed under well-regulated conditions in laboratories. These laws form the basis of a whole system of definitions. It is, therefore, appropriate that we should give attention to some aspects of definitions and the language and conventions used in them. Some aspects of definitions are also treated in Chapter 11 dealing with the language of chemistry and are relevant here; they will, therefore, not be repeated.

Variables   The variables used in the theory of physics differ from the variables which appear in mathematics. In mathematics we find that the variables come from the various *universes of discourse* which we discussed, for instance, from the set of all real numbers **R** (see Chapter 10). These numbers conform to the arithmetical rules of multiplication, division, addition and subtraction – which we all know from our schooldays.

There are three main kinds of variables in physics. The way in which they are used in physics differs markedly. They are as follows:

(1) The so-called 'dimensionless variables' which function just like the mathematical variables which we mentioned above, and which need no further discussion.

(2) The 'scalar variables' which are not just pure numbers, but consist of *two inseparable parts* which always occur together. These parts are:

☐ A *number* which refers to the numerical 'size' of the variable.
☐ A non-numerical part which is called its *dimension*.

The word scalar is derived from the Latin word *scala*, meaning 'ladder' and refers to the fact that the variable is 'ladder-like' or numerical.

(3) The *vector variables*, which are also not pure numbers, but which consist of *three inseparable parts* which always occur together. These parts are:

☐ A *number* which refers to the 'size' of the variable
☐ A non-numerical part which is called its *dimension*
☐ A non-numerical part which is called its *direction* which points into a specific direction in space.

The word vector is derived from the Latin word *vector*, meaning 'bearer' or 'carrier' – a bearer always carries something from one place to another, that is, he carries it in a certain direction and for a specific distance.

Each of the quantities (2) and (3) always functions as a *complete entity* in physics and their various parts can never be separated or 'forgotten'. Since these variables have such different attributes or properties, it is easy to imagine that the way in which they are used in calculations in physics will differ too. It will, therefore, come as no surprise to learn that each one of them has its own 'rules of arithmetic' which must be learnt.

Each of these variable types is written in a very specific way in physics. Conventionally, we write the number part first, which is then followed by the dimensional part, and then, if applicable, by the direction.

In the following text we'll first define what we mean by the term *dimension*. Then we'll discuss the measurement of these dimensions by means of another concept, called *unit*, referring only to scalar quantities for the moment. Towards the end of the chapter we'll give a brief discussion of vector quantities.

What is a dimension? We'll answer the question more fully below, but initially we say that a dimension is the answer to questions like: what do we have to measure to tell us how big something is? Or how long did it take us to do something? The answer to these questions is usually something like:

☐ *length* if we want to know the *size*
☐ *mass* if we want to know *how much* of something there is
☐ *time* if we want to know *how long* something took to do.

Once we know what we want to measure (for instance, the length of some object) we need to know which yardstick we'll need in order to make the actual measurement. The particular yardstick which we choose, is called the *unit* of the physical variable. We can thus replace the dimensional part of a variable by a part which indicates the *unit(s)* which we used for the measurement. In the sections below, we'll give more attention to these aspects.

☐ Dimensions and Units

The arithmetic of scalar quantities

We write a *scalar physical quantity or variable* as follows. Let r be such a physical variable – which consists of a *numerical part* and a *dimensional part*. We write it as follows:

$$r = \text{number}_1 \ \text{dimension}_1.$$

We have added the subscript 1 because we want to define more such variables below and we'll use these subscripts to distinguish between them.

The way in which we wrote the physical variable indicates quite clearly that it has two parts as explained above. This notation makes it clear that the numerical part is not multiplied by the dimensional part, but that the physical variable, or physical quantity, as it is also called, is a single entity, consisting of two parts which can never be separated and which always function together.

Let us look at another such physical quantity and let us call it s, with another numerical part and another dimensional part, or:

$$s = number_2 \; dimension_2.$$

Many things which we want to measure consist of products of dimensions, such as area which has dimensions of (length) × (length). The question is, what do we have to do when we multiply, divide, add and subtract two physical quantities, such as r and s? Obviously, the usual rules of arithmetic hold for the numerical parts. We also assume for the moment that a similar set of rules must also exist for the dimensional parts; for instance a rule which says: *What you do to the numerical part, you must also do to the dimensional part.* Let us try it by multiplying the variables r and s:

$$r \cdot s = [number_1 \; dimension_1] \cdot [number_2 \; dimension_2]$$
$$= [number_1 \cdot number_2][dimension_1 \cdot dimension_2]$$
$$= number_3 \; dimension_3$$

where we made the following substitutions:

$$number_3 = number_1 \cdot number_2$$
$$dimension_3 = dimension_1 \cdot dimension_2.$$

The multiplication of two such variables thus gives us a new numerical part as well as a new dimensional part and it seems as if our rule of multiplication works. We thus multiply the numerical parts together, as well as the dimensional parts. The same holds for division, where the numerical parts are divided, as well as the dimensional parts.

But what do we do in the case of the *addition* of two physical quantities? Let us add the physical quantities r and s:

$$r + s = [number_1 \; dimension_1] + [number_2 \; dimension_2]$$
$$= [number_1 + number_2][dimension_1 + dimension_2]$$
$$= [number_4][dimension_1 + dimension_2]$$

Handling the numerical part is not difficult: we are handling numbers and we know the normal rules of addition. But what about the dimensions? We did not add $dimension_1$ to $dimension_2$ because we know from experience that one cannot add one ox to one sheep to get two elephants. Although the argument is somewhat medieval and the two cases are not quite identical, it gives an idea of the problem of the addition of dissimilar dimensions: *we cannot add dissimilar dimensional parts*; the same holds for subtraction. The general rule which we stated above, namely, that the dimensional part is treated in exactly the same way as the numerical part during the four arithmetical operations, must thus be modified.

> It is clear that only physical quantities which have the same dimensions can be added and subtracted from one another.

This means that the addition of two physical quantities proceeds as follows:

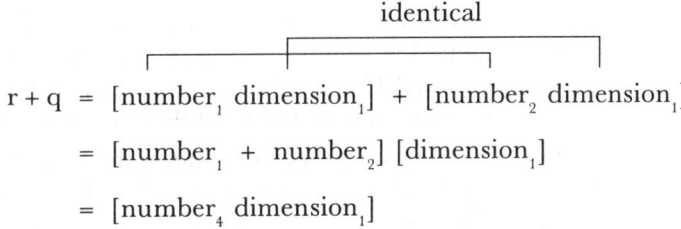

identical

$$r + q = [number_1 \; dimension_1] + [number_2 \; dimension_1]$$

$$= [number_1 + number_2] \; [dimension_1]$$

$$= [number_4 \; dimension_1]$$

It must be noted that we did not add the dimensional parts to get something like $\{2 \; (dimension)_1\}$ – we can understand this by again using our medieval analogy: adding one sheep and one sheep does not give two 2 sheep, but two sheep. The same holds for subtraction.

We thus have a rule for multiplication and division on the one hand, and another for addition and subtraction. These rules allow us to handle all scalar physical quantities which appear in physics.

Dimension   We have talked much about the concept of *dimension*, but what is it actually? We have partially answered the question in the introduction, but we'll add a few remarks here. When we want to know how big our bedroom is, then we measure its *length*; some reflection shows that its width which we measure next, is actually also a *length*. Then we calculate the *area* of the bedroom which everyone knows is (length × width) or, in terms of what we said above, the area is *length × length = length²*. Obviously, *length* is a dimension which we need in physics. What are the other dimensions which we use in physics?

The physicist has determined that there are just seven *base quantities* (and no more!) needed to describe the dimensions of all physical quantities which we find in physics. We have given three of them in the introduction, namely, *length, mass* and *time*. The others refer to the *electric current*, the *temperature*, the number of *distinct entities* in a piece of mass, and the *intensity of light*. Each of these base quantities is considered to have its own symbol and its own dimension. They are:

physical quantity	*symbol for quantity*
length	l
mass	m
time	t
electric current	I
thermodynamic temperature	T
amount of substance	n
luminous intensity	I_v

It is to be noted that each one of them has a symbol as indicated and that the symbol is always written in *italics* according to the rules of the International Union of Pure and Applied Chemistry (IUPAC) and the International Union of Pure and Applied Physics (IUPAP).

In the following discussion we use the internationally adopted convention that the inverse of a dimension is indicated by a negative exponent, for instance, $1/l = l^{-1}$; this is read *per length* or *length minus one*.

For instance, the dimensions of some physical quantities and the relevant abbreviations are as follows:

distance	length	l
speed	length per time	$l\, t^{-1}$
acceleration	length per time per time	$l\, t^{-2}$
momentum	mass length per time	$m\, l\, t^{-1}$

Analysing the dimensions involved in physical quantities and symbols can be very instructive and can identify any errors in equations and definitions. It is very important to realise that the dimension of an expression containing physical quantities which occur on the left of any equal sign (=) must be the same as that occurring on the right of the equal sign. Checking dimensions involved in equations may prevent errors, especially in examinations. Checking dimensions should actually be an automatic action on the part of a physicist, and the quicker one learns to do it, the easier will be the study of physics. There is an additional advantage: checking the dimensions involved in definitions and equations enhances the understanding of the theory, since it, as it were, forces one back to the fundamentals so that the interrelationship of definitions and theory becomes clear.

☐ Base Units

When we want to know how big our bedroom is, we measure the area, that is, we measure its width and its length and then we multiply the two together to give width × length; our dimensional analysis above shows that the dimensional symbol of area is l^2. This does not really help us, since we do not know what yardstick to use. There are many to choose from: we can use the Biblical *paces*, or the British *yard* or *foot* or *inch*, or the American *inch*, or the Cape *rood*, or the *metre*, etc. In fact, there is a bewildering variety of yardsticks to choose from. The same holds for any other physical quantity which we can think of. This causes confusion, and we need a yardstick or measuring *unit* for each of the seven base quantities which everybody around the world working in science, and in physics in particular, agrees to use. The emphasis is on the words 'agrees to use'.

The choice of measuring system was made by the Bureau International des Poids et Mesures, with headquarters in Paris. IUPAC and IUPAP agreed that they would accept this set of chosen yardsticks, or *base units* as they are called in scientific circles, as the basis of all measurements in physics and in chemistry. This choice is called the Système International d'Unités (abbreviated as SI). The units of these SI base quantities are as follows:

physical quantity	symbol for quantity	symbol of SI unit	SI unit
length	l	m	metre
mass	m	kg	kilogram
time	t	s	second
electric current	I	A	ampère
thermodynamic temperature	T	K	kelvin
amount of substance	n	n	mole
luminous intensity	I_v	cd	candela

The difference between the physical quantity, its symbol, the SI-base unit in which it is measured and the SI measuring unit should be clearly understood. It should be noted that the symbols for the SI units are by convention always written in upright type. Another convention is used, namely, that a space should be left between the number and the SI-units, as well as a space between the symbols of the units themselves. The symbols for physical quantities are always written in italics as we have done above in the tables.

As an example, consider the standard acceleration of free fall; it is written with a decimal point in the American and English parts of the world:

$$g_n = 9.806\ 65\ \text{m s}^{-2}.$$

The same expression is written with a decimal comma in Europe and in countries of the world which have strong ties to Europe:

$$g_n = 9,806\ 65\ \text{m s}^{-2}.$$

It should be noted that the symbol of the physical quantity itself is written in italics and the subscript n in upright type, because it denotes the word 'normal' which is not a symbol of a physical quantity. Another writing convention is also displayed here, namely, all numbers are grouped together in groups of three, counting to the left before the decimal point and to the right after the decimal point.

Much time has been spent by the international scientific community on the definitions of these SI base units. The definitions of the SI base units are given below, followed by the year in which they were ratified by the international community.[1] We give these definitions for the sake of completeness, to show how definitions are worded in a very exact way, and to furnish examples so that we can show how such definitions should be analysed and studied (even when one knows very little physics). We emphasise that they are not displayed to be learnt by heart, or to be understood as though we were teaching physics: their form and their language are the important things which must be noted (see chapter 6).

The SI base units

metre: The metre is a length of path travelled by light in vacuum during a time interval of 1/299 792 458 of a second. (1983)

kilogram: The kilogram is the unit of mass; it is equal to the mass of the international prototype of the kilogram. (1901)

second: The second is the duration of 9 192 631 770 periods of the radiation corresponding to the transition between the two hyperfine levels of the ground state of the caesium-133 atom. (1967)

ampere: The ampere is that constant current which, if maintained in two straight parallel conductors of infinite length, of negligible circular cross-section, and placed one metre apart in vacuum, would produce between these conductors a force equal to 2×10^{-7} newton per metre of length. (1948)

———— *continued* ————

---

1 The definitions of the SI base units are quoted from the IUPAC booklet: I Mills, T Cvitas, K Homann, N Kallay and K Kuchitsu, *Quantities, units and symbols in physical chemistry*. Published for the Union of Pure and Applied Chemistry, Oxford, by Blackwell Scientific Publications, Oxford, 1988.

— *continued* —

kelvin: The kelvin, unit of thermodynamic temperature, is the fraction 1/273.16 of the thermodynamic temperature of the triple point of water. (1967)

mole: The mole is the amount of substance of a system which contains as many elementary entities as there are atoms in 0.012 kilogram of carbon-12. When the mole is used the elementary entities must be specified and may be atoms, molecules, ions, electrons, other particles, or specified groups of such particles. (1971)

candela: The candela is the luminous intensity, in a given direction, of a source that emits monochromatic radiation of frequency 540×10^{12} hertz and that has a radiant intensity in that direction of (1/683) watt per steradian. (1979)

The meaning of the word *prototype*, which is used in the definition of the kilogram, is something like 'the first, or original or model from which anything is copied'. The kilogram is thus the standard against which all other masses are measured and other laboratories copy the standard kilogram for their own use.

It is clear that these definitions are not 'ancient', but that the international community is constantly refining them. They form part of the vocabulary of physics and they can be seen, as it were, to be part of a dictionary of physical terms and concepts. A dictionary is meant to be used when we do not know the meaning of a word or a term. A dictionary of physics functions in exactly the same way, and it is useful to construct one for personal use by entering all definitions encountered in any course consecutively in a notebook, especially after they have been studied and analysed in the way which we describe below.

This set of seven definitions looks (and is) rather formidable. The first impression is that of complexity: they are far more complex and exact than any definition or explanation of a word found in an ordinary dictionary. Every word counts, every word has a meaning, every number has an exact meaning. These definitions were very carefully phrased by the international scientific community so that there is no doubt whatsoever what we mean when we use the base units to measure our variables and to do our calculations. There are laboratories all over the world which use these standard definitions to produce secondary standards for science and technology which we can place in our laboratories and in our homes to measure.

The second thing that catches the eye, is the occurrence of rather strange numbers in some of the definitions, such as 1/299 792 458. The first idea that comes to mind upon reading them is that 'nobody in his right mind would pick such a number to put in a definition' that is, one would expect a 'round' number! The truth of the matter about these strange numbers is that they derive from units designed by previous generations of scientists and which were in use for a long time. A case in point is the second, which was defined a long time ago in terms of an earth-day. It was too much effort to introduce a new unit instead of the familiar second which generations of people knew and which we all use in our watches. The result is the strange number which links the old with the new.

The third thing which one notices is that all the base units, in spite of the fact that their definitions are very scientific and very exact, are units which were chosen by us: they are not 'natural units'. Nowhere in Nature do we find such units. We have chosen them for ourselves and we have agreed how they are to be measured. The exact definitions given merely specify how we measure them. For instance, the block of pure platinum metal which we use as the mass standard, the kilogram, is nothing more than a beautifully machined piece of metal which does not corrode (we hope) and which is hidden in a vault in a bank in Paris, France (so that it won't be stolen and so that it would not be exposed to the corrosive city atmosphere). We use it to determine the mass of everything else. The same is true of the mole, where we have arbitrarlily chosen a mass of 0.012 kilogram of a type of carbon called carbon–12 of which we count the atoms (it should be noted that the SI base quantity is called the mole, while its abbreviation is the mol).

Problem 1

Is the base unit of mass the only one which can be stolen, or are there others which ought to be hidden away in a bank in Paris?

Hint: Read every definition of the set of definitions very carefully!

Another thing which comes to the foreground when one reads these definitions carefully (and one should read all definitions very carefully, as explained in Chapter 6, which must be read together with these remarks), is that some of them depend on one or more of the others.

Problem 2

Carefully read the definition of the base unit called the *metre* and decide whether this unit depends upon any one of the other units or not. No knowledge of physics is needed to answer this question!

Problem 3

Carefully read this set of definitions and decide which of the definitions of the other base units do not depend upon any of the others. It is not necessary to understand the physics involved in order to answer this question!

Another point which we notice about these definitions is the number of other physics concepts which we must know before we can even attempt to understand the definitions, especially those which depend upon the three base units of Problem 3.

Problem 4

If we tell you that the definition of the watt depends upon the kilogram, the second and the metre, that the definition of the hertz depends upon the second, that the newton depends upon the kilogram, the metre and the second, set up a scheme showing the dependence of the other base units on the three base units of Problem 3. Again, no physics is needed to answer this problem.

The answer to this problem clearly shows that the SI base unit, the kelvin, as it were, stands alone: it is not used any further in defining any of the other base units. It also follows that the definition of the metre depends upon that of the second, while the definitions of the ampere and the candela, in turn depend upon that of the metre.

Problem 5

We said that some of these definitions are quite complex and that one needs to know quite a lot about physics to understand them properly. However, one does not need to know very much physics to make a list of all the concepts which are needed to understand a definition. Read the definition of the base unit *ampere* carefully and make a list of the concepts which one will have to look up in a physics course.

Problem 6

Pick out the main sentence or main statement of the definition of the *ampere*. If you find the main statement, then also specify the descriptive part and list the things which are described. Again, no physics is needed for an answer.

Problem 7

It can be argued that the definition of the ampere, as given above, has three serious omissions, where one or more words are left out in each of three places in the sentence. Can you determine which words they are, and where they should go?

Any definition, no matter how complicated, can and must be broken down and analysed in the same way so that it can be understood. The meaning of each term in the statement of a definition and in the accompanying descriptions must be found, and studied to see how it fits into the definition. It is especially important to see whether a definition is complete and does not make use of circular arguments (as was pointed out in the examples of Chapter 11). We do not pursue these matters any further, since that would lead us straight into the study of physics itself.

☐ Derived Units of the SI

It is inconvenient to specify the whole set of SI base units for every physical quantity which we measure. As an example, consider the physical quantity which we used above, namely, *energy*. The dimension of (energy) is the dimension (length^2 × mass × time^{-2}). If we translate this into the SI base units in which we measure mass, length and time we have the following:

units of (energy) = (metre^2 × kilogram × second^{-2})
$$\text{m}^2 \text{ kg s}^{-2}$$

It should be noted that it is conventional to list the abbreviations of the SI base units in any compound set of units in the order in which the definitions of the base units occur in the list and in the table as given above.

The product of units m^2 kg s^{-2} is definitely that of energy, but it is awkward to use and we prefer to abbreviate it by defining a SI derived unit for energy. In this case, the derived unit is the joule, with symbol J. The joule is defined in terms of the SI base units to be equal to m^2 kg s^{-2}.

There are two types of derived units in the SI, namely, those with special names and symbols, and those simply called 'other quantities' which are expressed in a mixture of the units with special names and the SI base units. There are 21 such special units in the SI, each with its special symbol and its definition in terms of the SI base units. We leave the subject here, since these derived units will be studied in physics courses.

We emphasise that every definition must be studied and analysed in the same fashion. It is also good practice to memorise the definitions of the derived units with special names, so that they become part of one's own vocabulary which can be used without having to consult the private dictionary of physical terms, or even the textbook.

☐ Visual Information in Physics

Visual information is very important in physics and is as much part of the language of physics as the words and the definitions. Various kinds of visual information are used, such as photographs and scale drawings of apparatus, perspective drawings of apparatus and molecules, projective drawings of all kinds of objects on paper, etc. We cannot go into detail here, but we mention again that any such visual information must be treated as part of the text and should be read and studied together with the text: text and visual information form one unit and can never be separated.

One often finds that definitions of terms or conventions used are only given in the form of figures or drawings; they are then not repeated in the text proper. These definitions play an important role in the text which follows such a drawing and every effort should be made to understand what message it is intended to bring across to the reader. One also finds that such definitions and conventions are relevant to some of the mathematics which is used in the text; every equation which is derived, should be checked against such conventions.

Right-handed co-ordinate system

An example of such practice, is the right-handed co-ordinate system of cartesian geometry which appears in many drawings in textbooks and which is generally used by the modern physics community. This co-ordinate system is depicted in **Figure 12.1**, which can be considered to be a definition of the system. An author might not even mention that the right-handed co-ordinate system is used to derive the equations: she assumes that the drawing speaks for itself. Older books sometimes make use of the left-handed cartesian co-ordinate system and care must be taken when comparing equations and rules which were derived in the two systems, since they can show differences (without either being wrong!).

Figure 12.1
The right-handed cartesian co-ordinate system, where the thumb of the right hand points in the direction of the positive z-axis, the forefinger to the positive x-axis and the middle finger to the positive y-axis.

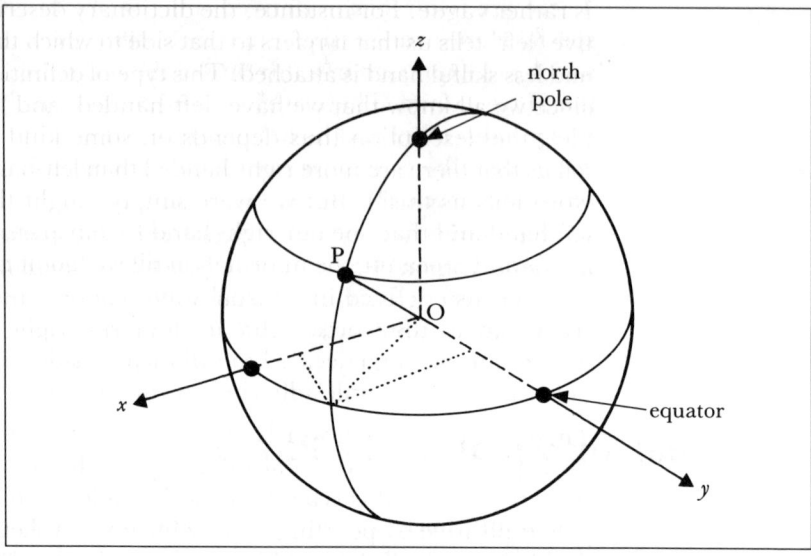

Problem 8

Decide whether the definition given in **Figure 12.2** is that of a left-handed or a right-handed co-ordinate sytem.

Figure 12.2
Figure which is used in Problem 8

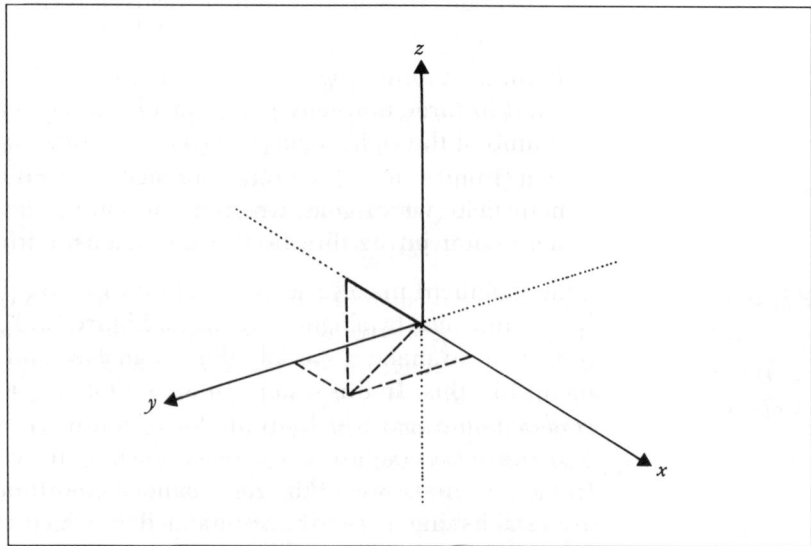

We have used the words 'right' and 'left' in the discussion above, with respect to our hands. The dictionary definition of such words is rather vague. For instance, the dictionary description of the adjective 'left' tells us that it refers to that side to which the normally weaker and less skilful hand is attached. This type of definition is rather 'weak', since we all know that we have 'left-handed' and 'right-handed people'; the description thus depends on some kind of statistics which tell us that there are more right-handed than left-handed people (which word tells us this?). But we were simply taught that this one is our left hand and that one our right hand by our parents pointing at our hands and repeating the information till we 'got it right'. These words are thus also defined in a visual way. There is nothing intrinsically 'right' about the word 'right' in the term 'right hand'. Somebody many centuries ago decided to call the one side 'right' and the other 'left', and we are still adhering to this perfectly arbitrary choice.

There are many such words and concepts in physics which are derived from essentially visual information. The words 'positive' and 'negative' are of this type. Somebody made a choice which electrical pole is positive by pointing at it by means of a sketch; somebody else decided to call the pole which turns to the north the north pole of a magnet. The words 'north' and 'south' are also of this type. We learn about them from visual information obtained elsewhere and they need not be defined anywhere in a physics textbook, but they are widely used in physics courses and one has to be on the lookout for them.

There is, for instance, a rule in physics which tells us something about the forces on currents in magnetic fields; it is called the *right-hand rule* and one finds descriptions like the following:

> Point the thumb, the forefinger and the middle fingers of the right hand in three mutually perpendicular directions. Then point the thumb of the right hand in the conventional direction of the current (from + to −), and the forefinger in the direction of the magnetic field (which goes from north to south); the middle finger will then point in the direction of the magnetic force.

This statement makes use of visual information (it may or may not be accompanied by a figure like that of **Figure 12.3**) and it is extremely important to make a special effort to analyse and understand statements like this. In this statement we are talking of three *mutually perpendicular directions* which are the three mutually perpendicular axes of the cartesian co-ordinate system which we have mentioned above. In fact, we make use of the right-handed co-ordinate system and we are establishing a set of conventions here which everybody follows. Nobody tells us how to point the fingers in the three mutually perpendicular directions, but there is only one way of doing this which is 'comfortable'. The word *mutual* in 'mutually perpendicular' tells

Figure 12.3
The right-hand rule
giving the direction
of the force
produced by an
electrical current in
a magnetic field.

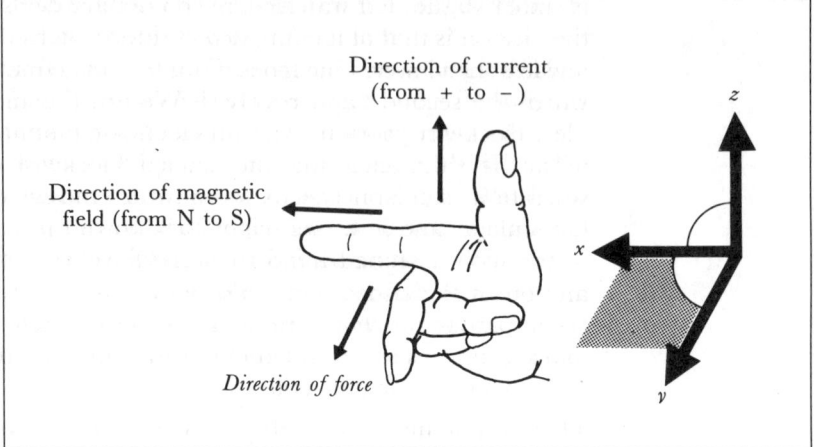

us that any two axes are perpendicular to the third (that is, the x- and the y-axes are each perpendicular to the z-axis, and the z- and the y- axes to the x-axis).

We are actually saying in this statement that we do the experiment in such a way that the electrical current flows towards the positive z-axis, and we place the magnet in such a way that the magnetic field points in the direction of the positive x-axis. We then say that if we do this, then the resultant force will point in the direction of the positive y-axis.

There are a few more conventions used in the statement which we must notice. The first one is the convention which we have, namely, that an electrical current flows from the + pole of the battery to the − pole through the external circuit (question: how does the current conventionally flow *inside* the battery?). This statement rests upon another convention, namely, that we abbreviate the word 'positive' by the + sign, and the word 'negative' by the − sign. The second convention, is that the direction of the external magnetic field is from the north pole of a magnet (the one that swings towards the north pole of the earth when a bar magnet hangs free from its centre) to the south pole. It is thus clear that this rather simple looking piece of prose is actually very compactly written and contains a large amount of information; it also needs much background information to understand properly. One must never learn such a piece of text by heart without really understanding all its aspects, otherwise the study of physics becomes 'mechanical' and unexciting, which it certainly is not!

Clockwise motion

We sometimes speak of a motion being *clockwise* around an axis which we imagine to be inserted in a body. This term refers to the direction of motion of the second hand (and thus that of both the other arms) of an old-fashioned watch. The conventional position for observing the motion is that of looking down at the watch which is with its face towards the viewer. One looks down (or along) the central axis around which the second hand revolves. We call this motion that we then see a clockwise motion. An anti-clockwise motion is just the reverse of this. It is thus clear that the concept 'clockwise' needs careful consideration, since one needs to know around which axis the motion takes place, and how one has to look down the axis at an imaginary watch with a second hand to judge the direction of the rotational motion of the body. This information is usually given in the text accompanying such a term, and one must search for it and one must make quite certain from which position one should view the motion around the particular axis.

There is one more thing about visual information which we meet in physics and with which we must be acquainted. This is the amount of *abstraction* which is mostly involved. It must never be forgotten that the physicist has a unique ability to strip a problem of all the 'inessential' things and to concentrate only upon those things which are important for the understanding of what is happening. The physicist wants to explain the phenomenon which is studied. The explanation takes place in the form of 'theory', which is built on an abstract model of reality. We do not like to deal with big bodies, so we say: let us think that this body is just as big as a point and that all its mass is condensed into the point. Then we apply the laws of point-mechanics to this particular point and its motion to see 'what we get'. An example of such an abstraction is shown in **Figure 12.4.**

Figure 12.4
How the physicist reduces the problem of two people carrying a pail of water. This shows that the components of the downward force experienced by the pail are divided amongst two people, a taller one and a smaller one. The larger the angle between their arms, the more they are 'pulled about' by the force.

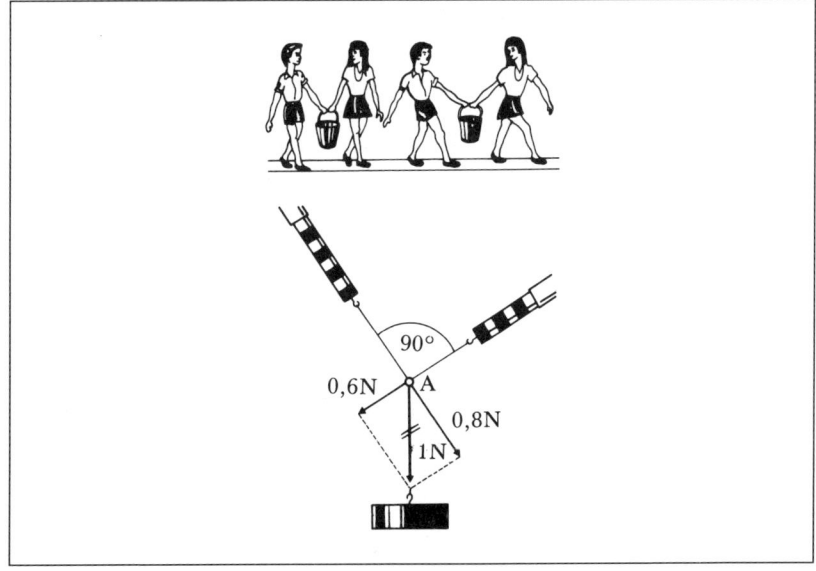

□ Graphs

Graphs are used extensively in physics – they appear everywhere and in many different forms. They form an integral part of the text and every graph displayed in a text is worth a careful study in which each part of the graph is carefully analysed. Anybody who is studying physics must also make a special effort to learn how to draw graphs from experimental and theoretical data, and how to interpret them. Most of the graphs which we find in physics are of the two-dimensional kind, using the x- and the y-axes of the cartesian co-ordinate system. There are other types of co-ordinate systems, but they are usually not found in introductory courses.

There are two main kinds of graphs in physics. The first type is obtained by plotting *experimental results* or *data*. The second type is drawn using *mathematical functions or equations* which were derived in the text. There are some things in common between these two types:

□ The co-ordinate axes which are used to plot the variables must take into account that the variable may have dimension(s) as explained above.

□ The actual plotting of the graph takes place after a table of the variation of the variable is obtained. The experimental type of graph obtains these data (note that *data* is a plural word!) from an experiment; the theoretical type obtains the table from a calculation.

□ The following plotting conventions are always used (and note how they depend upon words like *right*, *up* and *parallel*):

  – The *independent variable* is always plotted along the x-axis direction which is parallel to the lines on our notebook paper, with its positive part pointing to the right.

  – The *dependent variable* is always plotted along the y-axis which is taken to be perpendicular to the other axis and pointing 'upwards' to the top of the page.

The *independent variable* is always the one which is at our disposal: it is the one which we can change in the way we want. If we change the resistance of a resistor in an experiment, and then measure the current in the circuit, then the resistance (which we decide how to change) is the independent variable and is plotted along the x-axis. The current, which varies only when we change the resistance, is the dependent variable which is plotted along the y-axis.

Since we are only interested in the language aspects of physics, we cannot go into further detail here of how one actually plots these graphs and how one interprets them.

Reading a graph

How does one 'read' a graph? The first thing to remember is that every graph which is drawn from experimental data 'tells a story' about how a certain piece of matter (whether it is a resistor, or a gas which is enclosed in a glass container) behaves when we change a parameter which acts on it. If the points of the graph come from a function or relationship which we have derived from theory, then the story which it tells is how we think that this piece of matter *ought* to behave were we to change the parameter. It is the ideal of physics to develop the theory in such a way that the theoretical graph fits the experimental data as exactly as possible, or, put in another way, that the theoretical graph is well superimposed upon the experimental graph.

We look for the story behind the graph, that is, we must extract the story from the graph, just as we extract the story from the words, the sentences, the paragraphs and the chapters of a novel. This is done step-wise:

☐ In the first place, one must look at the axes to determine 'what is plotted against what', that is, what the dependent and independent variables are.

☐ It is very important to determine which *units* are involved along both axes since they are used in all calculations: a physical variable always comes attached to its unit. There are various conventions or ways in which the units can be displayed on the axes of a graph and they will be met in physics courses. The important point, however, is that they are always displayed.

☐ The *scales* along the axes must be found and noted, since this gives one the feeling for the order of magnitudes (the 'sizes') of the variables occurring in the graph (they are, in general, different along the two axes). There are also various correct ways in which the scales can be attached to the axes, but we leave that to the physicist to explain. The important point is that they are always displayed along the axes.

☐ The *shape of the curve* must be noted. There are two main types, each one having its own characteristics which must be noted:

    – The *linear type* (which is the one that the physicist and the chemist really appreciate!). An example of this type is shown in **Figure 12.5**.

    – The *non-linear type* which can come in many shapes. Such a typical curve is shown in the graph of **Figure 12.6**.

Figure 12.5
An example of a
linear graph. The
dotted line is the
extrapolated part of
the line, that is, into
the region where we
do not have any
experimental data
points, but where
we believe that the
line would have
appeared if we had
made such
measurements.

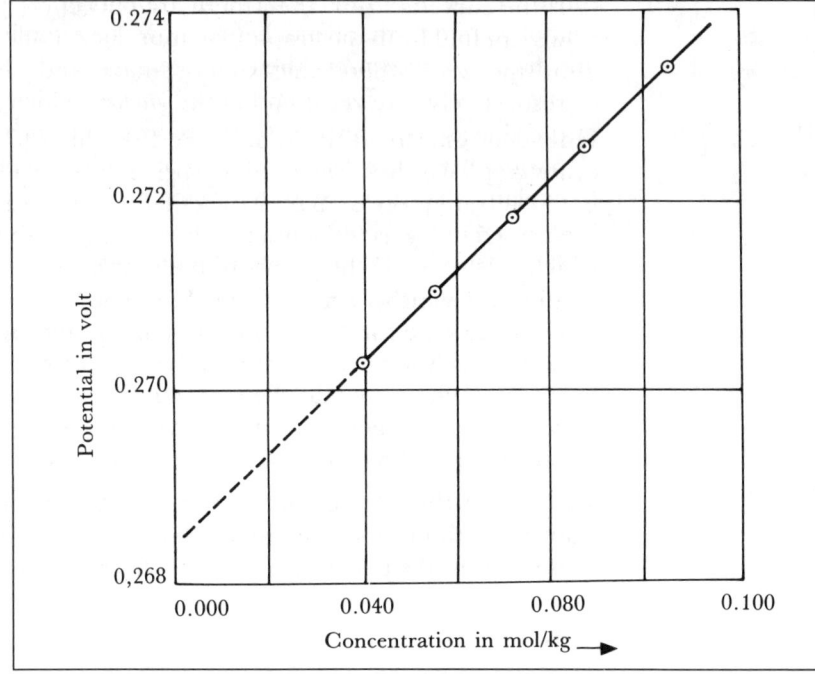

Figure 12.6
A typical example of
a non-linear curve.

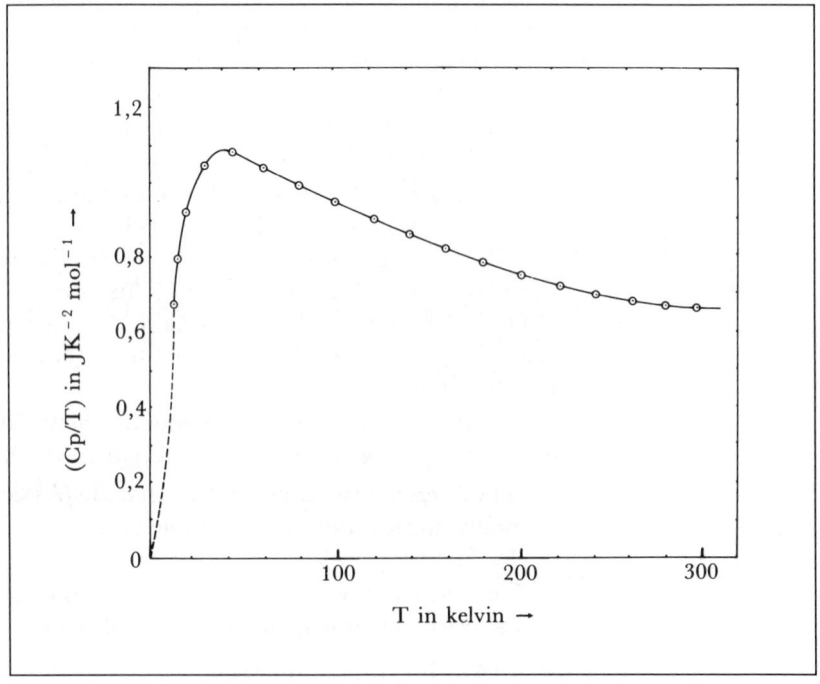

☐ If the curve is *linear*, then there are only three parameters which we can find from the graph: one must look at the two cut-off points at the *axes* where the x-co-ordinate and the y-co-ordinate, respectively, are zero; and at the *gradient* which tells us how steep the line is with respect to the x-axis (the meaning of the three parameters are learnt in school mathematics and we cannot explain what they are in this chapter). These three parameters tell us very much about the piece of matter and how we see it from the theoretical point of view. Sometimes one also wants the *area* enclosed under the curve between two x-co-ordinates.

☐ If the curve is *non-linear*, then we have to look at points where the curve reaches its *maximum* or *minimum*, or where *inflections* occur; the co-ordinates of these points mean something in terms of the background theory. We must also look at the co-ordinates of the points where the curve cuts the axes.

☐ If two curves are plotted on one graph, then we are sometimes interested in the *co-ordinates of the points where they intersect* and we must look for the meaning of these numbers.

☐ Vector Quantities

We defined a vector above as a quantity that has three attributes, namely size (a number), units (dimension) and direction. We can thus write a vector quantity **r** as follows:

$$\mathbf{r} = \text{number}_1 \ \text{dimension}_1 \ \text{direction}_1$$

A vector is mostly printed in a physics textbook in such a way that it is clearly distinguished from a scalar quantity. IUPAC insists that the symbols of vector quantities must be printed in **bold type** as we have done above, but the physics community proper has many different ways of indicating such a symbol. This sometimes involves special symbols such as placing a little arrow over the symbol for the vector quantity.

An example of a statement regarding a vector quantity is:

She is cycling northwards with a velocity of 15 km h^{-1} against a sideways headwind having a velocity of 10 km h^{-1} blowing from the north-west.

It is clear that we have two different velocity vectors here, namely:

$\mathbf{v}_1$ = 15 km h^{-1} to the north

$\mathbf{v}_2$ = 10 km h^{-1} to south-east.

Problem 9

The use of the prepositions in the expressions 'to the south-east' and 'from the north-west' must be noted. Do you agree that both expressions refer to the same direction?

Figure 12.7
The four different directions N, S, E, and W on earth

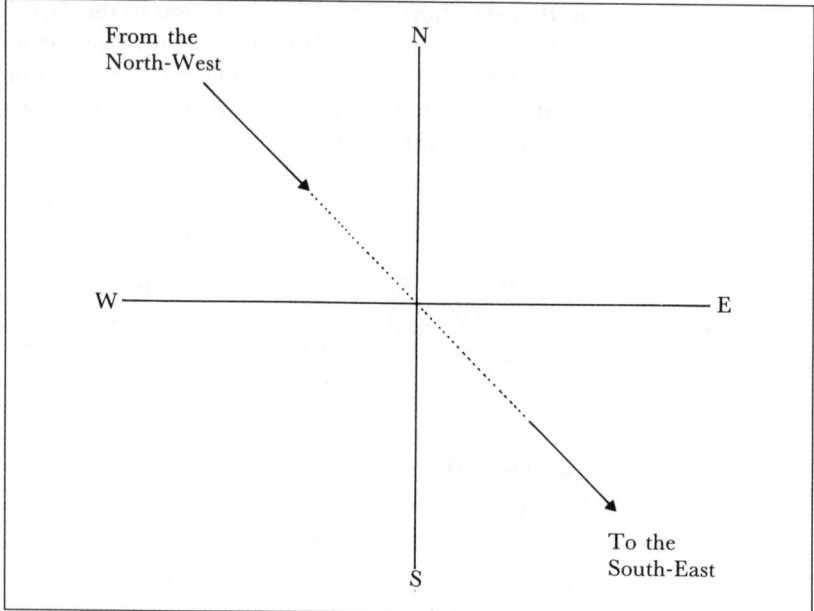

If we now want to add these two vectors $\mathbf{v}_1$ and $\mathbf{v}_2$, what happens? We know that we can only add units which are identical (which is the case here). But what about the directions? If we go through the addition sum (similar to the one we discussed for scalars), we find:

$$\mathbf{v}_1 + \mathbf{v}_2 = \text{(velocity of 15 km h}^{-1}\text{ to the north)} + \text{(velocity of 10 km h}^{-1}\text{ to south east)}.$$

It is clear that we need a completely new 'arithmetic' or algebra to handle such quantities and we give an example below to make the reader conversant with the language and symbolism which is found in physics courses.

The notation given above for a vector quantity is very cumbersome and inconvenient. The mathematician has a way of indicating vector quantities which the physics community took over. Instead of giving names like East or North to the four ends of the planar cartesian system of axes (see **Figure 12.7**), two *unit vectors* are attached to the centre of the two axes, pointing along the axes. This is illustrated in **Figure 12.8(a)**. These unit vectors are the *direction indicators* of the two perpendicular axes: they give the direction of the axes. The product of the co-ordinate of a point P on an axis and the unit vector of that axis indicates the vector from the origin to the point P.

These unit vectors are variously abbreviated by symbols, but the physicist prefers to call the one pointing along the x-direction **i**, while the one along the y-direction is called **j**; for three-dimensional problems, the one along the z-direction is called **k**. It is worthwhile learning these symbols for the three cartesian unit vectors by heart, since they are used everywhere. Attention is drawn to the fact that some textbooks and lecturers prefer their own set of unit vector symbols. (It does not really matter which set of symbols are used, as long as they are clearly defined, either in a statement or by means of a sketch.) One finds equivalent symbols such as:

$$
\begin{array}{ccc}
\mathbf{i} & |\,i> & |\,1> \\
\mathbf{j} & |\,j> & |\,2> \\
\mathbf{k} & |\,k> & |\,3>
\end{array}
$$

The most popular symbols for introductory courses are **i**, **j** and **k**.

The vector from the origin to the point P on the x-axis with co-ordinate $x = 5$, is then indicated by the expression:

$$\mathbf{P} = 5\mathbf{i} \text{ dimension}$$

where we have simply indicated the dimensional part of the vector by adding the word 'dimension', without specifying it further.

The vector from the origin to the point Q on the y-axis with co-ordinate $y = 3$ is given by the expression:

$$\mathbf{Q} = 3\mathbf{j} \text{ dimension.}$$

In the same way, is the vector from the origin to the point S with co-ordinates $x = 5$ and $y = 3$ [2] given by the sum of the two previous expressions:

$$\mathbf{S} = \mathbf{P} + \mathbf{Q} = (5\mathbf{i} + 3\mathbf{j}) \text{ dimension.}$$

---

2 Conventionally, we use the notation that the point has the co-ordinate pair (or more simply, co-ordinates) (5,3). The two co-ordinates are enclosed in a bracket pair, separated by a comma. The first co-ordinate conventionally refers to the x-co-ordinate, while the second refers to the y-co-ordinate. The notation often used for vectors in physics courses is that the numerical part of a vector is written as a row vector, such as (5 3), without any commas between the numbers. The unit vectors are then written as a column vector:

$$\begin{pmatrix} i \\ j \end{pmatrix}.$$

The vector **S** is then written as the dot multiplication of a row vector by a column vector:

$$(5\ 3) \cdot \begin{pmatrix} i \\ j \end{pmatrix} = 5\,i + 3\,j$$

It is clear that a peculiar type of multiplication is involved.

This is geometrically illustrated in **Figure 12.8(b)**. There are also other algebraic methods to do the same calculation, but for this, one needs a new and very interesting algebra called *vector algebra* – which is one of the most useful algebras to know, but we cannot give justice to its language here!

Figure 12.8
(a) The unit vectors **i** and **j** in the cartesian plane. These unit vectors point, respectively, along the x- and the y-axes (which are shown as dotted lines to emphasise the unit vectors)
(b) The way in which the physicist writes the vectors **S**, **P** and **Q** discussed in the text

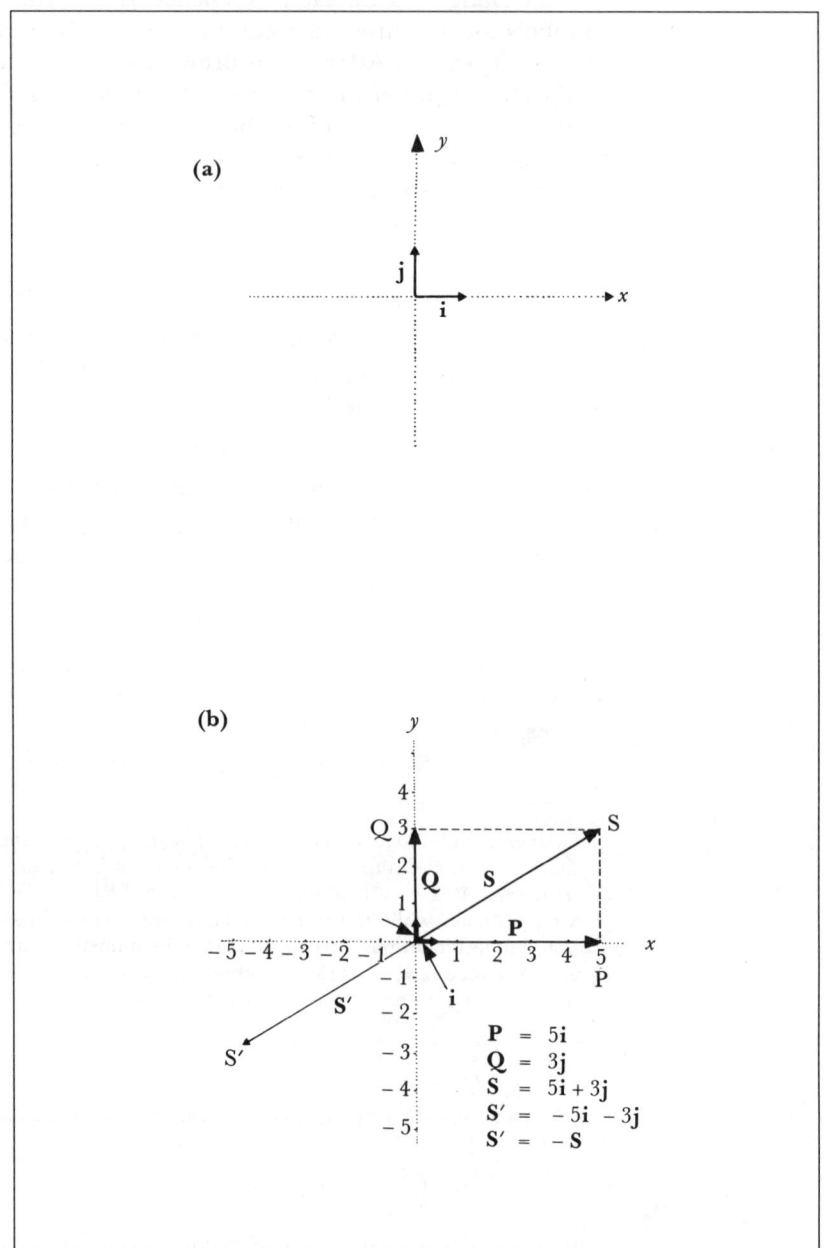

Problem 10

Write down the vector pointing from the origin to the point S' with co-ordinates $x = -5$ and $y = -3$. Is this vector S' the negative of the vector S, that is, does it point exactly in the direction opposite to that of the vector S?

The answer to problem 10 means that we can add these two vectors S and S' which originate in the same point (the origin with co-ordinates (0,0) in this case) so that we'll obtain the *zero vector* (note: not the number zero, but the zero vector):

$$S + S' = (5i + 3j) \text{ dimension} + (-5i - 3j) \text{ dimension}$$
$$= [(5 - 5) i + (3 - 3)j] \text{ dimension}$$
$$= (0i + 0j) \text{ dimension}$$
$$= O \text{ dimension.}$$

We emphasise again that there is a world of difference between the concept *zero* and that of the *zero vector*, and they should never be confused. We gave an illustration to show how vectors are added; the multiplication of vectors, however, will take us too far into physics proper.

☐ Answers

Problem 1

The kilogram, the unit of mass is the only one which relates directly to a physical object (if we lose the kilogram, then we lose our mass unit). As far as we know, nobody has ever tried to steal it! All the other SI base units are derived from *measurements* – which can be repeated on any appropriate piece of matter, provided we have the correct apparatus which is sensitive enough to do it. We can let light travel in vacuum for that fraction of a second required by the definition and use that distance as our base unit for length. We can count the correct number of periods of radiation and then say that the time elapsed is one second. The same holds for all the other SI base units. We wish to draw the attention to the fact that it is not at all easy to do these measurements, and very sensitive instruments in specially-designed laboratories are needed.

Problem 2

It is quite clear when one reads the definition of the concept *metre* that it can only be understood, if the concept *second* is understood: the metre is defined in a way which depends upon the definition of the *second* (which part of the definition tells us that this is so?).

Problem 3

The definitions of the *second* (which we answered in Problem 2), the *kilogram*, and the *kelvin* do not depend upon the definition of any one of the others, since their definitions do not involve any one of the others. These three base units thus seem to be the 'most fundamental' of the seven defined base units of the SI.

Problem 4

We can make the following diagram to show their interrelationship when we read the set of definitions and the extra information which we have given very carefully:

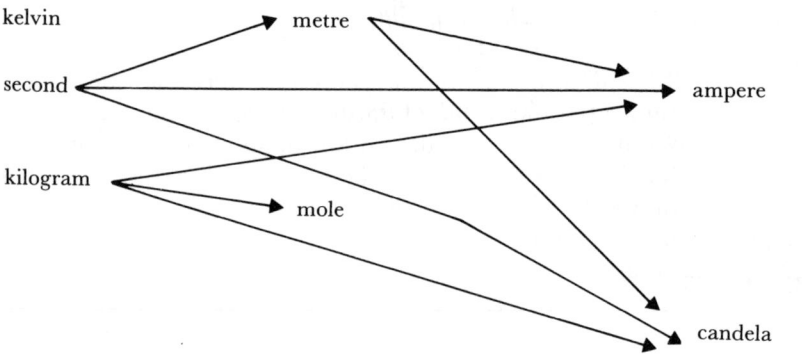

Problem 5

The concepts we have to look up in our dictionary of physics (or in the textbook) if we want to understand the definition of the *ampere* are as follows, in the order of appearance:

current	constant current
conductor	straight conductor
	straight conductor of infinite length
	straight conductor having almost no diameter
	two parallel conductors
diameter	circular cross-section of a thin straight rod
metre	
vacuum	
force	newton
	newton metre^{-1}
	force between two parallel conductors
	carrying a constant current

Problem 6

The main statement of the definition is the part that tells us that a current of one ampere through the conductors produces a force of 2×10^{-7} newton per metre of the length of the conductors.

The rest of the statement merely specifies the type of conductor which we are interested in, what it looks like, how long it is, and where it is placed.

The definition can thus be interpreted to say the following, where both parts must be read together:

The definition itself:

The ampere is that constant current which must be maintained in two straight parallel conductors to produce a force between them which is equal to 2×10^{-7} newton per metre of length.

The description of the conductors:

The two straight parallel conductors must be:
☐ infinitely long
☐ circular in diameter
☐ so thin that their diameter is almost zero
☐ placed in a vacuum
☐ one metre apart.

Problem 7

In the first place, the definition speaks only of a *current*; its only qualifying adjective is the word *constant*. One has no idea of what sort of current it refers to. Thinking back to our school physics, we find that there is more than one current type in physics: direct electrical current, alternating electrical current, water current, air current, ion current, etc. Which of these does the SI unit refer to? The current actually referred to, is the direct electrical current. We *infer* (which means: logically deduce that this must be the case) from the presence of other words which are used in the definition, namely 'conductor', 'vacuum' and 'force', but without a knowledge of physics one would never know which type of current is meant. Definitions must be complete and exact to be of full use. (This must be remembered in the examinations too: half a definition is actually no definition at all!)

In the second place, we have to ask ourselves the question: where is the current maintained? In the one conductor, or the other, or in both? If the current is in both, how is it divided amongst the conductors? In such a fashion that the sum of the currents in the two parallel conductors is equal to one ampere? Or does each conductor carry a constant current of one ampere? The definition is thus not clear on this point. The physicist assures us that the last possibility is indeed the

— *continued* —

---
continued
---

correct one: each conductor carries a constant curent of one ampere. It would have been better if the definition had read : .... maintained in *each of* two straight parallel conductors ....'.

In the third place, we can ask the question: What experiences the force? One of the conductors, the other conductor, or both? If both experience the force, how are the 2×10^{-7} newton per metre of length divided between the two conductors? The physicist again assures us that each of the conductors experiences a force of 2×10^{-7} newton per metre of length. Again, the sentence would have been more clear reading as follows: '... between each of these conductors a force ...'.

When one knows the theory of electricity, then, perhaps, it is not strictly necessary to insert these 'missing' words, since one would then know the intent of the definition. However, from a point of view of the completeness condition of a definition, these inserted words are definitely needed.

Problem 8

The definition is that of a left-handed co-ordinate system, since the thumb of the left hand is directed towards the positive z-axis, the forefinger of the left hand to the positive x-axis, and the middle finger of the left hand to the positive y-axis. Only the axes should be considered for the definition; the rest of the information on the figure may be relevant for another discussion, but not for the definition of the co-ordinate system.

Problem 9

Make a little drawing of two perpendicular axes and put the names of the four different directions (N, S, E and W) on the ends of the axes and place a little arrow to indicate 'to the south-east' and another for 'from the north-west'. It is thus clear that the *direction* indicated by the two phrases is the same. This is shown in **Figure 12.7**. The problem where the 'observer', (who views the direction) should be standing, raises its ugly head again in this case, and it ought to be carefully considered.

Problem 10

When the two vectors are drawn on a set of axes with the appropriate dimension, then we find that they point to directly-opposing directions.

13

Visual Information in Science

☐ Introduction

Visual or non-textual material also forms part of the language of science. It is, therefore, necessary to know how to deal with such information. There are many types of such visual information, some of which are almost self-evident and need no teaching. Other types are rather involved and special skills are required to utilise the information to the full. We cannot give attention to all the various types in only one chapter. We have, therefore, selected some important aspects of visual information for inclusion. Some other aspects of visual information are also covered in previous chapters dealing with the individual disciplines.

☐ Pictorial Information Signs, or Icons

Definition

> The dictionary definition of an *icon* is 'a figure; an image; in the Russian and Greek Orthodox Churches a figure representing Christ, or a saint in painting or mosaic'.

The purpose of such an icon is to tell a story or to provide an image of the person who is revered for people who cannot read or to those whose belief will be reinforced by such images or icons. A icon thus transmits pictorial or non-verbal information to a viewer which is essential for him to know. The *central figure* is drawn on a (wooden) *board* which is usually square or oval, and is surrounded by a *border*. The images are usually coloured paintings or even carvings in relief.

299

The pictorial information signs (we may even call them icons) used in our modern world are quite different from the religious ones, but they have all the design features in common with them, such as:

☐ the *image* (which has now become quite *stylised* and does not show any detail)

☐ the *board* (which may be made of wood, metal, plastic, cardboard, paper, or even a computer screen); the shape is mostly circular, triangular or square; even the *shape* of the board also transmits information

☐ The *border* which is mostly reduced to a single coloured band encircling the central image

☐ They also convey a *non-textual and non-verbal message* to the viewer which gives direct access to important information which the viewer *must take heed of or react to*.

The modern icon-like sign also displays some text under certain conditions, but ideally it should no do so, since the icon should be designed in such a way that it conveys all its information in a visual and non-textual fashion so that its message can cross any language and cultural barrier in a way that cannot be misread. It is quite an effort to design such a 'fool-proof' icon-like sign!

Everybody knows and uses these pictorial information signs or icons all the time – and we obey them without question if we want to have a long and healthy life! The signs which we are talking about, are the *traffic icons* or *traffic signs* (on land, sea and air). We must learn these road signs to obtain our driver's licences and we must take heed of them when we drive on the roads, when we walk anywhere near or on a road, or when we cycle anywhere. We do not reproduce such signs here, since they are so well-known. Other signs give information in hotels, such as 'here is the bathroom', or 'this is the staircase', 'run this way in case of fire'. And just as our lives are protected when everybody adheres to the instructions and information provided by the road signs, so can our lives (and those of our fellow laboratory workers) be saved by knowing the international laboratory signs.

Icon-like signs occur on the walls of laboratories, the doors and corridors leading into them, on glass and other containers containing (dangerous) chemicals, on trains and tankers transporting chemicals, in plant rooms in laboratories, on the doors of laboratory store-rooms. They should be known to the user of any one of these facilities and, more important: their message should be understood and heeded. They are there to protect the life of any person working in the area for which they are 'responsible'. The message of an icon should never be ignored. In some countries it may be an offense to ignore such scientific icon-like signs, or to neglect to put them up where they are needed.

The most important types are *warning signs, information signs, prohibitive signs, mandatory signs, safety signs* and *computer-screen icons*.

Warning signs

The *warning signs* are *triangular* in shape and tell us about the presence of something which is hazardous to our lives. The danger which they warn of is not a possible danger, but a real danger. They should, therefore, be immediately heeded and the prescribed protective measures taken. Such signs indicate the presence of *radioactive material*, the presence of *inflammable material*, open electrical wires carrying a dangerously *high voltage and/or current* which could produce a shock, *laser radiation* which could damage the eyes and the skin, the presence of *non-ionising radiation* such as X-rays, etc.

The design of these signs is very stylised and makes use of a set of images which represents the dangerous situation. Every effort should be made to learn at least those displayed in one's own working area. Sheets explaining the meanings of relevant signs are usually prominently displayed in any laboratory; such information sheets should be carefully studied. if an unknown sign occurs anywhere, it is usually good practice to enquire about its meaning and message – it might save one's life! Some important signs which belong to this type are displayed in **Figure 13.1**.

Figure 13.1
Some representative examples of warning signs

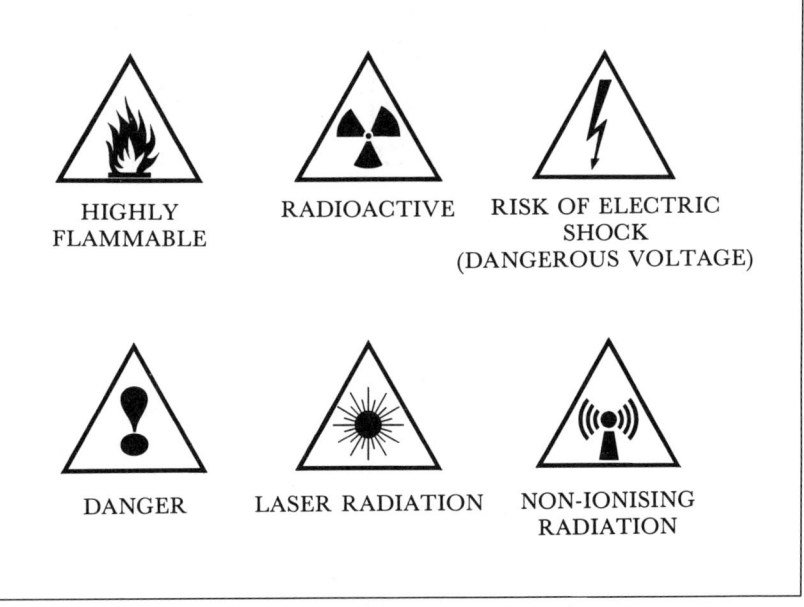

Information signs

These *information signs* are either *square* or *diamond-shaped* and are mostly intended to be put on the labels of containers which contain dangerous chemicals. The diamond-shaped signs are put on tankers and other vehicles which transport chemicals (even traffic policemen and firemen are supposed to know them!).

The image on the sign usually conveys a very clear picture of the danger inherent in the chemical about which it warns. The signs are clear enough so that even farmers using such chemical should have no doubt about them. Such signs indicate the possibility that the chemical may be *explosive* (such as fine zinc powder, or TNT), that it is of the *oxidizing type* (like hydrogen peroxide), that it is *corrosive* (like an acid, such as hydrochloric acid or hydroflouric acid), that it is highly *flammable* (such as ether, or acetone), that it is a harmful *irritant* to the eyes or the mucous membranes of the mouth, nose or lungs, or even to the skin in general (such as ammonia), or that it is *toxic* or *poisonous* (such as arsenic pentoxide or potassium cyanide). Some representative signs are shown in **Figure 13.2**.

Some other such signs are those indicating the presence of glassware in a container which may easily break, or the double arrows indicated which side of the container ought to be the upper side.

Figure 13.2
Some representative signs of the information type which are placed on chemical containers such as bottles, gas cylinders, drums, tankers, etc

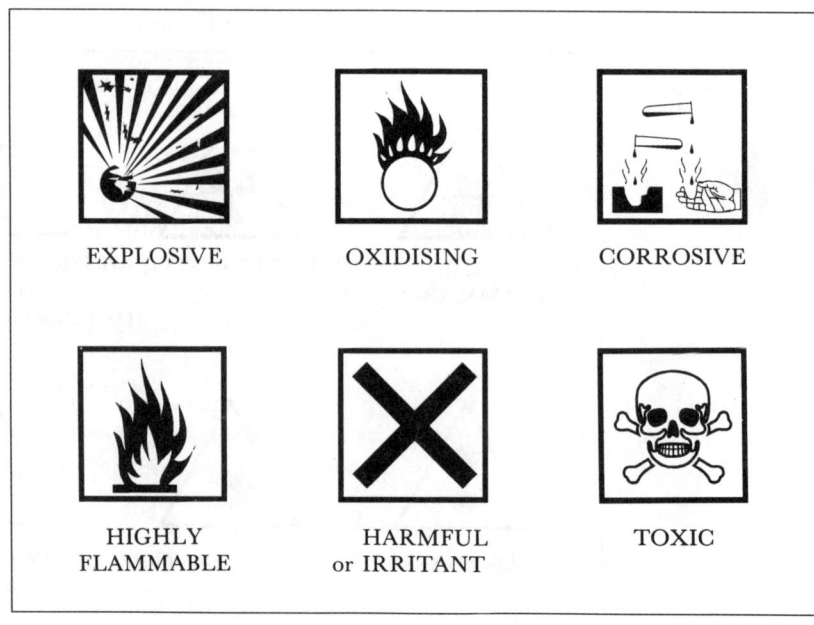

| EXPLOSIVE | OXIDISING | CORROSIVE |
| HIGHLY FLAMMABLE | HARMFUL or IRRITANT | TOXIC |

There are other types of signs too, for instance, the way to correctly connect the electrical wires to an electrical wall plug is usually shown in the form of a series of drawings of icon-like simplicity.

Prohibition signs

Figure 13.3
A typical prohibition sign, telling us not to smoke under any circumstances

These signs are similar to some prohibitory traffic signs, such as 'do not overtake in the next 100 metres', or 'do not stop here'. Such scientific signs prohibit certain behaviour in laboratories, store rooms for chemicals, etc. They instruct us 'not to have any naked flames around' (fire or explosion hazard), 'when this catches fire, do not extinguish with water' (fire or explosion hazard when fire is doused with water – for instance a fire caused by the metal potassium), 'do not smoke', etc.

These prohibition signs carry a band diagonally across an image. This band conveys the message: 'do not . . .'. One such sign is displayed in **Figure 13.3**.

Mandatory signs

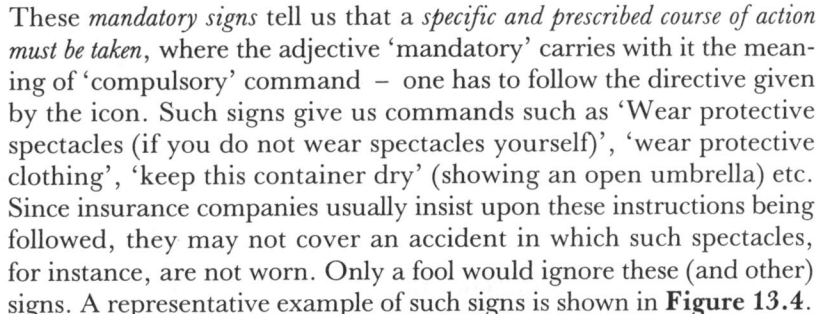

Figure 13.4
An example of a mandatory sign giving the instruction 'wear protective spectacles'

These *mandatory signs* tell us that a *specific and prescribed course of action must be taken*, where the adjective 'mandatory' carries with it the meaning of 'compulsory' command – one has to follow the directive given by the icon. Such signs give us commands such as 'Wear protective spectacles (if you do not wear spectacles yourself)', 'wear protective clothing', 'keep this container dry' (showing an open umbrella) etc. Since insurance companies usually insist upon these instructions being followed, they may not cover an accident in which such spectacles, for instance, are not worn. Only a fool would ignore these (and other) signs. A representative example of such signs is shown in **Figure 13.4**.

Safety signs

Figure 13.5
A representative example of a safety sign

These *safety signs* tell us about where safety or first-aid boxes may be found, as well as to inform us about other safety conditions, or they can give us safety information in general. One of the most useful is the sign indicating where the first-aid basket may be found. Such a representative safety sign is displayed in **Figure 13.5**.

Computer screen icons or icon verbs

These are icons appearing in certain blocks on the computer screen and they function as the imperatives of certain verbs which instruct the computer to do something. When the *cursor* (the little sign that shows the user where the next letter will appear on the screen) is moved to such an icon block and the *enter* or acceptance key is pressed, the computer automatically executes the desired instruction.

Such *icon verbs* are often used in computer aided instruction (educational) programs and give instructions to the computer 'to proceed to the next page of text', 'to return to the beginning of a section', etc. There is unfortunately, no internationally-agreed upon set of icon verbs which the writers of such programs are obliged to use. One of the most common is the 'hand icon' with (or without the pointing forefinger) which indicates the instruction to the computer to proceed with the display of the next page.

☐ Colour Codes

Science is full of colour codes and it is worthwhile for anybody who uses a laboratory to know some of the most important colour codes. Most of these colour codes are now 'internationalised' so that everybody tends to use the same colour code for the same thing. These colour codes tend to use the basic colours, as well as striped colours, although colour mixtures are also found.

There are colour codes for *electrical wires*:

☐ *red* for the *positive wire* of direct current
☐ *black* for the negative wire of direct current
☐ *brown* for the *live wire* of alternating current
☐ *yellow with green stripe* for the *earth wire* of alternating current
☐ *blue* for the *neutral wire* of alternating current.

These colour codes are extremely important, since a wrong electrical connection may be dangerous not only to oneself, but may also destroy expensive apparatus.

The manufacturers of complex apparatus with many wires may make use of private colour codes to facilitate the tracing of faults and the connection of the wires.

There are colour codes for the commonly-used *laboratory gases*, where both the gas cylinders, as well as the gas pipes which lead to the benches (where applicable) are coloured. The most important colour codes for gases according to SABS 019-1985 are:

☐ french grey for nitrogen
☐ black for oxygen
☐ middle brown for helium gas
☐ golden yellow for chlorine gas.

It is important to note such colour codes, since they may prevent costly and dangerous mistakes.

☐ Electrical Signs

These are symbols which were designed to represent certain electrical components in electrical and electronic circuit diagrams. Such signs are very useful since they contract a word into a visual sign which is very easily readable in a complex circuit. For instance, the signs for some electrical and electronic components are given in **Figure 13.6**. It is usually of great help to make a dictionary list of signs and their meanings encountered during a course. Every effort should be made to memorise such symbols, since they are also part of the language of the science.

Figure 13.6
Abbreviating signs for some electrical and electronic components

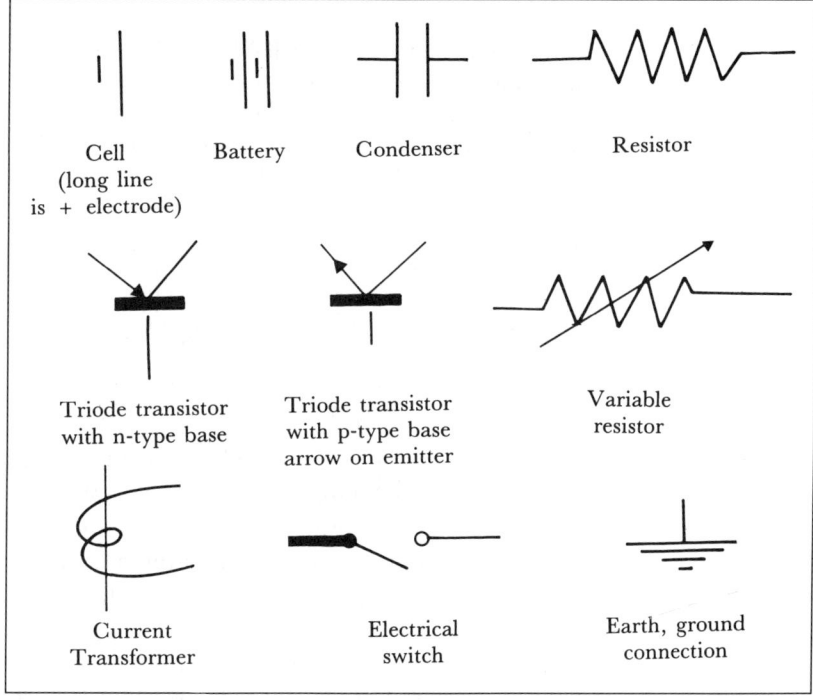

The layout of any circuit diagram is actually a prime example of very compact visual information; this will be dealt with in more advanced physics courses.

☐ Flow Charts

It is often possible to break any complicated process down into a sequence of simple actions. Breaking any such process down into a sequence of steps ensures that nothing is overlooked or left out and that all the steps have been identified. Making such a breakdown can thus be a very important learning aid and we recommend that the procedure be used as often as possible. This is especially important in the experimental sciences where a set of complicated actions have to be undertaken in a precise sequence to do an experiment correctly so that meaningful conclusions may be drawn.

We'll illustrate it with a breakdown of the complicated process of getting up in the morning; obviously, we cannot show all the details, but we give enough to give a general idea how to proceed. The writing conventions which are in use are as follows:

☐ The *start and end points of the action or process are shown as oval blocks*, containing the words START and STOP, respectively.

☐ The *actions undertaken*, which are all imperatives, such as SHOWER and SHAVE, are shown in *square blocks*.

☐ The *decisions to be made* are all shown in *diamond-shaped blocks* which display the words YES and NO at their sides, indicating the possible answers to the single and clearly-phrased question which is asked inside the block, the answers determining the path which we take through the flow diagram.

☐ the consecutive actions/decisions are all written down in sequence underneath one another, linked by an appropriate line.

A simplified 'getting-up' flow chart is displayed in **Figure 13.7** and the reader is advised to study this diagram very carefully, noting how this type of visual information can convey a massive amount of information. It is worthwhile to puzzle a bit over the function of the decision blocks and to determine just how they do their jobs and why they are there (they allow one to cover all possibilities, for instance, the case of married persons, as well as single persons). The particular flow chart is a bit long to display in sequence down a single page and it is 'folded' for the sake of convenience.

It is emphasised that the construction of this type of flow chart ought to be mastered, since it can make life easier, especially during practical sessions. The whole complex sequence of operations known as the *qualitative analysis of the metal cations* in chemistry depends upon a flow chart such as this. The system of identification depends upon such a flow chart. Computer programs are also written after a problem has been broken down into a sequence of steps and a flow chart (or flow diagram) has been constructed. It can even be used in a fault finding situation to make certain that all the right questions are asked and all decisions are correctly taken. It also functions excellently for certain kind of logic problems occurring in chemistry, as well as for the naming of organic chemical compounds according to the so-called IUPAC-scheme of nomenclature (for more particulars, see Chapter 11). Students often experience difficulties comprehending these schemes of analysis or programming since they have never studied the construction and use of flow charts. It is easy to apply this type of analysis to other real-life situations to get more confidence.

Figure 13.7
A simplified flow chart describing the process of getting up and getting started in the morning. The various writing conventions used are explained in the text

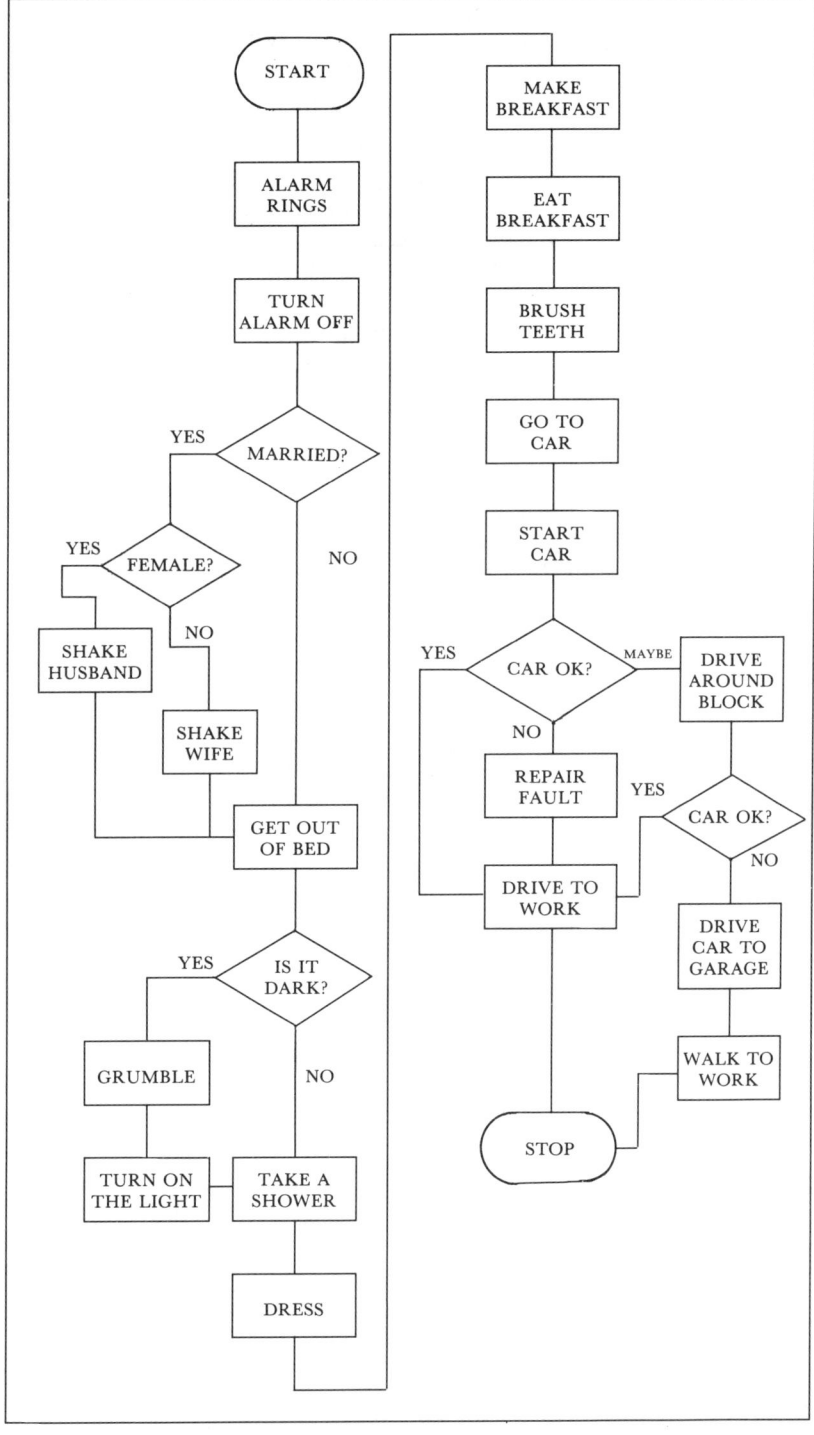

Figure 13.8
A scheme to analyse
a given *pure salt* to
establish whether it
is a salt of lead(II)
or mercury(I)

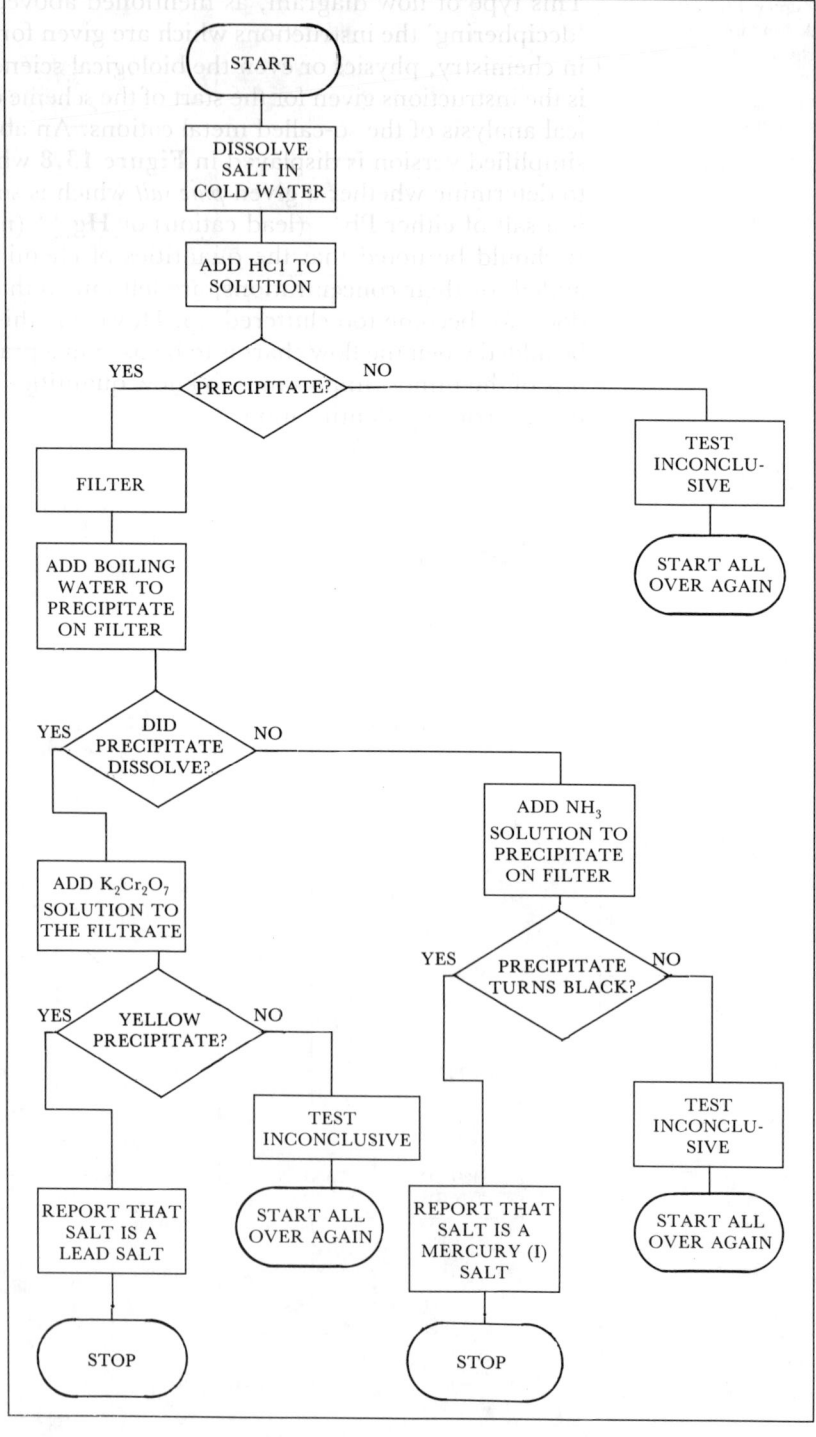

This type of flow diagram, as mentioned above, is of great help in 'deciphering' the instructions which are given for many experiments in chemistry, physics or even the biological sciences. A case in point is the instructions given for the start of the scheme of qualitative chemical analysis of the so-called metal cations. An abbreviated and very simplified version is displayed in **Figure 13.8** which will enable one to determine whether a given *pure salt* which is soluble in cold water is a salt of either Pb^{++} (lead cation) or Hg_2^{++} (mercury (I) cation). It should be noted that the quantities of chemicals which must be added, or their concentrations, are left out so that the flow diagram does not become too cluttered-up. However, this information must be added when the flow chart is to be used in a practical session, since it is of the utmost importance to know quantities and concentrations for the correct identification.